Haptic Technology - Intelligent Approach to Future Man-Machine Interaction

Edited by Ahmad Hoirul Basori, Sharaf J. Malebary and Omar M. Barukab

Published in London, United Kingdom

IntechOpen

Supporting open minds since 2005

Haptic Technology - Intelligent Approach to Future Man-Machine Interaction
http://dx.doi.org/10.5772/intechopen.94628
Edited by Ahmad Hoirul Basori, Sharaf J. Malebary and Omar M. Barukab

Contributors
Pablo Sanchez-Sanchez, José Daniel Castro-Díaz, Alejandro Gutierrez-Giles, Javier Pliego-Jiménez, Vladimír Hain, Roman Hajtmanek, Dušan Kočlík, Clement Onime, Peter J. James Van Den Broek, James O. Uhomoibhi, Mattia Santachiara, Ahmad Hoirul Basori, Omar M. Barukab, Sharaf Jameel Malebary, Andi Besse Firdausiah Mansur

First published in London, United Kingdom, 2022 by IntechOpen
IntechOpen is the global imprint of INTECHOPEN LIMITED, registered in England and Wales, registration number: 11086078, 5 Princes Gate Court, London, SW7 2QJ, United Kingdom
Printed in Croatia

British Library Cataloguing-in-Publication Data
A catalogue record for this book is available from the British Library

Additional hard and PDF copies can be obtained from orders@intechopen.com

Haptic Technology - Intelligent Approach to Future Man-Machine Interaction
Edited by Ahmad Hoirul Basori, Sharaf J. Malebary and Omar M. Barukab
p. cm.
Print ISBN 978-1-80355-066-4
Online ISBN 978-1-80355-067-1
eBook (PDF) ISBN 978-1-80355-068-8

For EU product safety concerns:
IN TECH d.o.o., Prolaz Marije Krucifikse Kozulić 3, 51000 Rijeka, Croatia, info@intechopen.com or visit our website at intechopen.com.

Meet the editors

Ahmad Hoirul Basori is a full professor in the Faculty of Computing and Information Technology, King Abdulaziz University, Saudi Arabia. Prior to this, he was an assistant and associate professor at Universiti Teknologi Malaysia where he obtained a Ph.D. in Computer Graphics in 2011. Dr. Basori is an editorial board member for several international journals and has published more than 100 articles. He is also a member of the Institute of Electrical and Electronics Engineers (IEEE), International Association for Computing Machinery's Special Interest Group on Computer Graphics and Interactive Techniques (ACM SIGGRAPH), and International Association of Engineers (IAENG) and a senior member of International Association of Computer Science and Information Technology (IACSIT). His research interests include computer graphics, facial animation, cloth simulation, medical visualization, haptic interaction, man-machine interaction, and robotics.

Omar M. Barukab obtained a Ph.D. in Computer Engineering from the College of Engineering, Florida Institute of Technology, the USA in 1999. Currently, he is a full professor and dean of the Faculty of Computing and Information Technology, King Abdulaziz University, Saudi Arabia. He is an active researcher and teacher. His research interests include logic, mobile agents, cryptography, data mining, and information security and auditing.

Dr. Sharaf J. Malebary received his Ph.D. in Computer Science and Engineering from the University of South Carolina, USA in 2018. He is a senior member of the Institute of Electrical and Electronics Engineers (IEEE) He has held many managerial positions and currently serves as vice-dean of development at the Deanship of Admission and Registration, and vice-dean of graduate studies and scientific research at the Faculty of Computing and Information Technology, King Abdulaziz University. His research interests include antennas and radiation patterns, drones and unmanned aerial vehicles (UAVs), sensor modalities, machine learning, deep learning, and network security.

Contents

Preface

Haptic technology has become the key to bringing realism into a virtual world. The idea of an immersive world will affect the physical interaction between the user and the virtual object. Haptics or the sense of touch is the engagement of senses beyond sight and hearing. Haptics has various models, ranging from simple and small sensors to complicated devices. From mobile phones to exoskeletons, forced haptic feedback is very expensive and complex.

This book provides an overview of haptic research for various modalities in human-machine interaction. For example, haptic technology can be used for multiple medical applications and help doctors or patients to fulfill their needs. Haptic technology is expected to gain more flexibility, accuracy, and reality during the interaction. The patient might feel the sensation of a doctor's existence through their skin via various methods of haptic technology such as vibration, mid-air haptic feedback, and forced feedback. The latest technology, known as mid-air haptics, can generate tangible feelings on the human skin. The Airborne Ultrasound Tactile Display (AUTD) is one of the mid-air haptics with great medical or entertainment potential. Moreover, haptics might be helpful for feeling the surfaces and contours of heritage buildings. Haptic technology opens the possibility of heritage sensation to the next level by delivering to the user the feeling of being inside an old building. Haptics has also become the evolutionary interface for mixed reality applications where physical and virtual realities can be integrated as one interface. In the robotics field, haptics is useful for improving robotic perception and vision. Haptics helps improve robot navigation by giving the robot a better sense of its surroundings.

The book consists of five chapters divided into three main sections that cover the robotic interface of haptic technology, haptics for medical purposes, and haptics for digital heritage.

Section 1 begins with Chapter 1, which is an introductory chapter that discusses various intelligent approaches toward future human-machine interaction. Chapter 2 discusses the collaboration of robot and haptic feedback for enhancing robotic perception. Chapter 3 presents an evolutionary user interface of haptic and mixed reality.

Section 2 includes Chapter 4, which examines how haptic technology is used for medical simulation.

Section 3 includes Chapter 5, which discusses the possibility of using haptics to enhance the presence of heritage building surface sensation to augment digital heritage documentation.

Ahmad Hoirul Basori, Omar M. Barukab and Sharaf Jameel Malebary
King Abdulaziz University,
Faculty of Computing and Information Technology in Rabigh,
Department of Information Technology,
Rabigh, Saudi Arabia

Section 1

Robotic Interface of Haptic Technology

Chapter 1

Introductory Chapter: Intelligent Approach to Future Man - Machine Interaction

Ahmad Hoirul Basori and Andi Besse Firdausiah Mansur

1. Introduction

Industry 4.0 has changed the interaction model between humans and machines. They have defined the "machine" as an automatic process that includes hardware and software [1]. Furthermore, they also discussed the elements of controller modules and administration. Industry 4.0 also brings industrial change from analogue toward digital solutions [2, 3]. Human–machine communication is also enhanced in the high capacity of computer networks [4]. The integration of a Cyber-physical system (CPS), Internet of Things (IoT) and cloud network would be feasible by an advanced network that has high-speed internet access [5].

2. Future man-machine interaction

The evolution of man–machine interaction has come up with some pillars of innovative technology-focused areas such as big data, robots, self-driving cars, and Augmented and Mixed Reality [1]. Big data analytics helps give direction to complex criteria of decision-making problems. For example, robotics might better perceive their object detection and recognition environment. The other development of human-computer interaction is driven by mobile device and interface device growth. Bieller mentioned that a voice-guided user interface becomes accustomed to people's lives, and it's been predicted in the next 5 years, it will adopt for more than 80% of technology usage [6].

Furthermore, Augmented Reality (AR)/Virtual Reality (VR) is also widely used because VR gear price is less price, and AR technology can run on smartphones smoothly. Industry 4.0 also strive for the development of autonomous vehicle (AV). Stoma et al. [7] classify the level of automation for an autonomous vehicle into several groups, such as level 0 (without automation), level 1 (assistive driving), level 2 (partial automation), level 3 (Conditional Automation), level 4 (advanced automation) and level 5 (full automation). Level 5 means that car can navigate automatically without any human intervention. It can accomplish all driving requirements in all circumstances [7–9]. Furthermore, incorporating haptic technology into machines makes interaction more realistic because haptic can give people a sensation of the object in the virtual environment [10]. So they can sense the presence of entities like they felt in the real world.

The chapters introduce and demonstrate the future man–machine interaction in intelligent ways. The man–machine relations relied on developing devices that drive people to change their way of interacting with a machine. Industry 4.0 has incorporated more Artificial intelligence for future interaction and robotic involvement and evolutionary interfaces such as Haptic, AR or VR. The book is concentrated on researchers, industry professionals, graduate students and academicians who need a solution or knowledge for an innovative interaction between humans and machines.

Acknowledgements

This work was supported by the Deanship of Scientific Research (DSR), King Abdulaziz University, Jeddah Saudi Arabia. The authors, therefore, gratefully acknowledge the DSR technical and financial support.

Author details

Ahmad Hoirul Basori* and Andi Besse Firdausiah Mansur
Faculty of Computing and Information Technology in Rabigh, Makkah, Saudi Arabia

*Address all correspondence to: abasori@kau.edu.sa

References

[1] Nardo M, Forino D, Murino T. The evolution of man–machine interaction: The role of human in industry 4.0 paradigm. Production & Manufacturing Research. 2020;**8**(1):20-34. DOI: 10.1080/21693277.2020.1737592

[2] Frank G, Santos Dalenogare L, Ayala NF. Industry 4.0 technologies: Implementation patterns in manufacturing companies. International Journal of Production Economics. 2019;**210**(4):15-26

[3] Kumar R, Haleem A, Garg SK, Singh RK. Automated guided vehicle configurations in flexible manufacturing systems: A comparative study. International Journal of Industrial and Systems Engineering. 2015;**21**(2): 207. DOI: 10.1504/IJISE.2015.071510

[4] Gu Y, Liu Q. Research on the application of the internet of things in reverse logistics information management. Journal of Industrial Engineering and Management. 2013;**6**(4):963-973

[5] Sony M. Industry 4.0 and lean management: A proposed integration model and research propositions. Production & Manufacturing Research. 2018;**6**(1):416-432

[6] Bieller E. What's the Future of Human-Computer Interaction (HCI)?. 2021. Available from: https://careerfoundry.com/en/blog/ui-design/the-future-of-human-computer-interaction-whats-next/ [Accessed: 1 March 2022]

[7] Stoma M, Dudziak A, Caban J, Dro'zdziel P. The future of autonomous vehicles in the opinion of automotive market users. Energies. 2021;**14**:4777. DOI: 10.3390/en14164777

[8] Gavanas N. Autonomous road vehicles: Challenges for urban planning in European cities. Urban Science. 2019;**3**:61

[9] Siqueira Silva D, Csiszár C, Földes D. Autonomous vehicles and urban space management. Scientific Journal of Silesian University of Technology. Series Transport. 2021;**110**:169-181

[10] Sreelakshmi M, Subash TD. Haptic technology: A comprehensive review on its applications and future prospects. Materials Today: Proceedings. 2017;**4** (2, Part B):4182-4187

Chapter 2

Robot Perception Based on Vision and Haptic Feedback for Fighting the COVID-19 Pandemic

Ahmad Hoirul Basori, Omar M. Barukab,
Sharaf Jameel Malebary and Andi Besse Firdausiah Mansur

Abstract

The robot perception can be enhanced further through visual and haptic to give more impression. This chapter aims to combine vision and haptic for the robot navigation during tracing their movement. The pandemic has striven humans to do direct contact; therefore, an alternative using the robot as delivery tools is assumed to be one of solution. As the initial experiment has been shown in the previous section, the deviation of angle is quite low and the success rate of arriving at the destination is also quite high around 76%. Future work can be enhanced by improving the success rate by monitoring the robot track closely.

Keywords: robot perception, vision, haptic, navigation

1. Introduction

The pandemic of COVID-19 has affected human life in general. Many prevention actions are used to prevent the virus from the spread, including maintaining social distance and wearing a facemask. To ensure the safety of people, robots are used for controlling the COVID-19 patient bed using the Arduino robot [1]. It is designed to make less contact with patients so reduced the chance of infection. The Coronavirus spreads via the saliva droplets or nose liquid when the sick person is coughing or sneezing [2]. Due to the high rate of infection, mobile robots can be an alternative solution to reduce contact with patients [2]. Robots for support service is one of the solutions for maintaining society awareness toward the virus spread. The mobile robot's utility during the pandemic might vary, such as delivery service, population awareness, and disinfection facilities [2]. This chapter aims to provide new robot insight through deep learning vision and haptic that can augment the robot's response toward their environment.

2. Related works

Artificial Intelligence (AI) and robotics are valuable resources for helping the patient treatment, doctors, nurses, and other front-line staff. Intelligent robots can perform good service for a particular task when it is planned and designed better [3, 4]. However, due to the high cost and complexity of the technology, not all country affords to adapt this approach [2]. Furthermore, another researcher tried to utilize a nursing robot for patient monitoring and medicine consumption according to the medication schedule. The other robot known as Lio-A is a robot with a multi-functional arm that has the capability of human-robot interaction and personal care assistant (**Figures 1** and **2**) [5].

The Lio-A robot has a visual and audio sensor for receiving the command, while laser and ultrasound for navigation and surrounding monitoring. It also has a mechanical sensor for handling the task given to them. Lio-A has the capability for autonomous action by having automatic navigation and recharging [5]. The other researcher also uses robots and realistic virtual reality to enhance interaction between humans and machines. The interaction can be in a gaming-based system, Brain-computer interface, or 3D simulation [6–14]. Virtual navigation using augmented reality or sensors is also helpful for the robot to achieve the desired direction according to the path that set up for them [15–28]. Machine learning

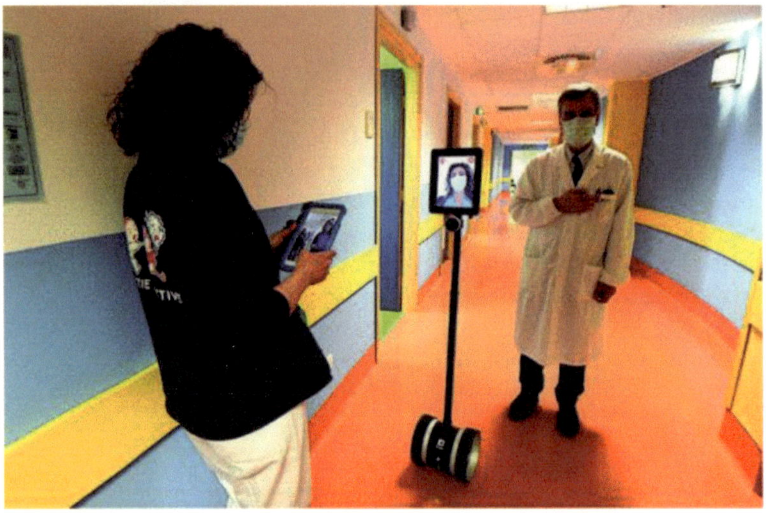

Figure 1.
Robot consultation [2].

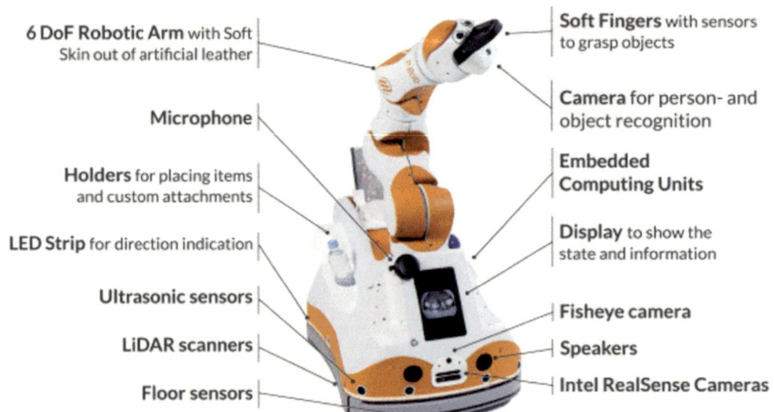

Figure 2.
Lio-A robot for a personal assistant [5].

algorithms, such as reinforcement learning also involve managing the crowd behavior of pedestrians, so it will be beneficial for robotic navigation later [29–33]. In addition, the Kinect camera with its capability for gesture tracking is also beneficial for medical applications along with robotics for helping COVID-19 patients, doctors, and nurses [34–41].

3. Methodology

The visual perception of the robot can be enhanced through a special camera, such as a Kinect depth camera, that can provide a depth image stream as input. This image will be useful for the navigation of the robot due to its capability of providing a 3D image of the object.

4. Materials

There is three main hardware that involved during experiments:

- Wheel robot using Arduino board Uno3, refer to **Figure 3**.
- Kinect camera attached to the robot body.
- Laptop for processing.

The Kinect camera will stream the object in front of the robot and send it to the pc, once pc received the image. It will be continued by processing the image whether the image is one of its goals or obstacles. If obstacle then the robot needs to avoid the object, while if it is a goal, the robot needs to grasp the object using the gripper, a detailed methodology is shown in **Figure 4**.

Figure 3.
System setup.

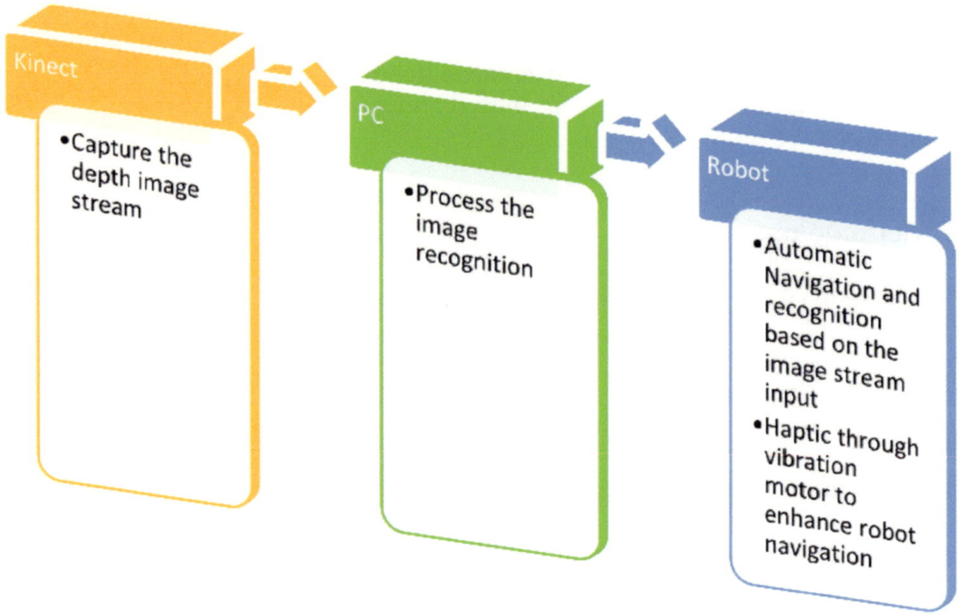

Figure 4.
System methodology.

5. Result and discussion

The initial testing using the Kinect camera is tracking the user skeleton. It can track the user's movement, especially hand gestures that can be interpreted as command control. The tracked skeleton is shown as a fragmented line that is imposed on the human body, as shown in **Figure 5**. The parts of the body, such as the arm, body, and head skeleton, are tracked in real time. Later this body part will be used as reference control to manipulate the robot.

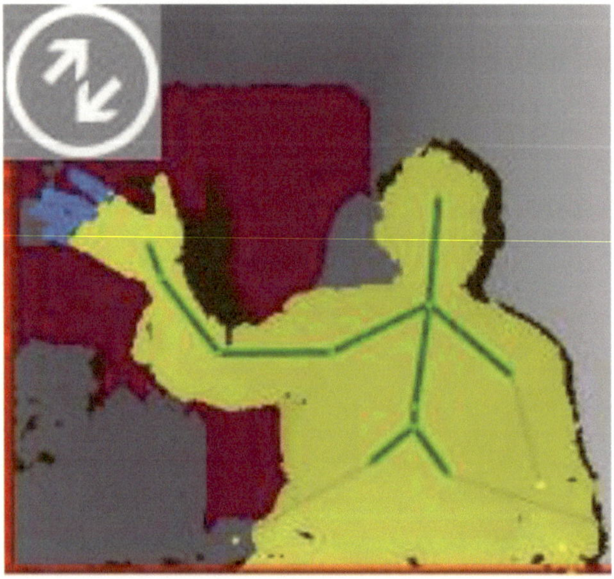

Figure 5.
Skeleton tracking.

Figure 6 shows how the depth image stream of the object streamed into pc for further process. While the depth image with a recognized person is shown in **Figure 7**.

Figure 6.
Depth image stream.

Figure 7.
Person recognition.

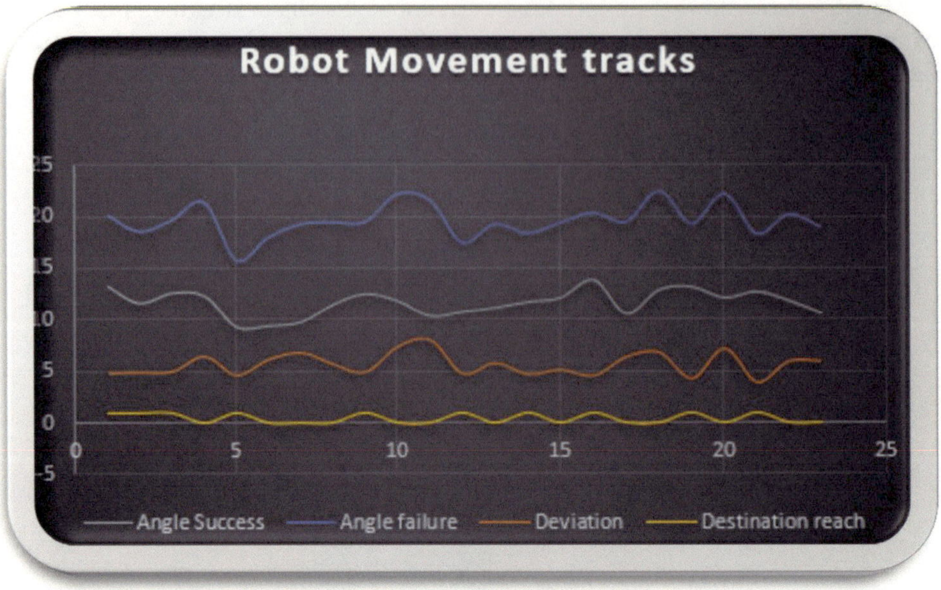

Figure 8.
Robot movement tracks.

The process of person recognition with pre-train model is quite interesting; the robot can recognize the person by receiving the information from the main process inside the PC. The haptic device is used for helping the robot on avoiding obstacle by sending the vibration as an alert signal. So, the robot can turn left or right to avoid the obstacle. We did several experiments with robotic movement, as shown in **Figure 8**.

6. Conclusions

Visual and haptic enhancement of robotic perception is very important for the success rate of the robot task. In this chapter, we present a combination of haptic with the vision to enhance the robot navigation during performing the delivery task to the user. Robot delivery is one of the essential keys during the pandemic of COVID-19 to avoid direct contact between humans. As the initial experiment has been shown in the previous section, the deviation of angle is quite low and the success rate of arriving at the destination is also quite high around 76%. Future work can be enhanced by improving the success rate by monitoring the robot track closely.

Acknowledgements

This work was supported by the Deanship of Scientific Research (DSR), King Abdulaziz University, Jeddah Saudi Arabia. The authors, therefore, gratefully acknowledge the DSR technical and financial support.

Author details

Ahmad Hoirul Basori*, Omar M. Barukab, Sharaf Jameel Malebary
and Andi Besse Firdausiah Mansur
Faculty of Computing and Information Technology in Rabigh, Makkah,
Saudi Arabia

*Address all correspondence to: abasori@kau.edu.sa

References

[1] Hadi HA. Line Follower Robot Arduino (using robot to control Patient bed who was infected with Covid-19 Virus). In: 2020 4th International Symposium on Multidisciplinary Studies and Innovative Technologies (ISMSIT). 2020. Istanbul, Turkey: IEEE;

[2] Cardona M, Cortez F, Palacios A, Cerros K. Mobile robots application against covid-19 pandemic. In: 2020 IEEE ANDESCON. Quito, Ecuador: IEEE; 2020.

[3] Hussain K, Wang X, Omar Z, Elnour M, Ming Y. Robotics and artificial intelligence applications in manage and control of COVID-19 pandemic. In: 2021 International Conference on Computer, Control and Robotics (ICCCR). Shanghai, China. 2021

[4] Lay-Ekuakille A, Chiffi C, Celesti A, Rahman MZU, Singh SP. Infrared monitoring of oxygenation process generated by robotic verticalization in bedridden people. IEEE Sensors Journal. 2021;**21**(13):14426-14433

[5] Mišeikis J, Caroni P, Duchamp P, Gasser A, Marko R, Mišeikienė N, et al. Lio-A personal robot assistant for human-robot interaction and care applications. IEEE Robotics and Automation Letters. 2020;**5**(4): 5339-5346

[6] Abdullasim N, Basori AH, Salam S, Bade A. Velocity perception: Collision handling technique for agent avoidance behavior. Indonesian Journal of Electrical Engineering. 2013; **11**:2264-2270

[7] Afif FN, Basori AH. Orientation control for indoor virtual landmarks based on hybrid-based markerless augmented reality. Procedia—Social and Behavioral Sciences. 2013;**97**:648-655

[8] Afif FN, Basori AH, Saari N. Vision-based tracking technology for augmented reality: A survey. International Journal of Interactive Digital Media. 2013;**1**(1):46-49

[9] Ahmed MA, Basori AH. The influence of beta signal toward emotion classification for facial expression control through EEG sensors. Procedia—Social and Behavioral Sciences. 2013;**97**:730-736

[10] AlJahdali HMA, Basori AH. Emotional contagion driven of parent-child's agents in crowd during panic situation. International Journal of Computer Science and Network Security. 2019;**19**(1):1-7

[11] Basori AH. End-effector wheeled robotic arm gaming prototype for upper limb coordination control in home-based therapy. TELKOMNIKA Telecommunication, Computing, Electronics and Control. 2020;**18**(4):2080-2086

[12] Basori AH, Aljahdali H. Abdullah OSJIJoACS, applications. Forced-Driven Wet Cloth Simulation based on External Physical Dynamism. 2017;**8**:572-581

[13] Basori AH, Aljahdali HJCE. Applications. TOU-AR:Touchable Interface for Interactive Interaction in Augmented Reality Environment. 2017;**6**:45-50

[14] Ahmed MA, Basori AH, Saari N. Brain computer interface control for facial animation. International Journal of Interactive Digital Media. 2013;**1**(2):58-61

[15] Albaqami NN, Allehaibi KH, Basori AH. Augmenting pilgrim experience and safety with geo-location way finding and mobile augmented reality. International Journal of

Computer Science and Network Security. 2018;**18**(2):23-32

[16] Allehaibi KHS, Basori AH, Albaqami NN. B-COV:Bio-inspired virtual interaction for 3D articulated robotic arm for post-stroke rehabilitation during pandemic of COVID-19. International Journal of Computer Science and Network Security. 2021;**21**(2):110-119

[17] Arpit J, Abhinav S, Jianwu W, Mangey R. Use of AI, Robotics, and Modern Tools to Fight Covid-19. Use of AI, Robotics, and Modern Tools to Fight Covid-19. River Publishers; 2021. pp. ii-xxx

[18] Basori AH. Emotion walking for humanoid avatars using brain signals. International Journal of Advanced Robotic Systems. 2013;**10**(1):29

[19] Basori AH. HapAR: Handy intelligent multimodal haptic and audio-based mobile AR navigation for the visually impaired. In: Paiva S, editor. Technological Trends in Improved Mobility of the Visually Impaired. Cham: Springer International Publishing; 2020. pp. 319-334

[20] Basori AH, AlJahdali HMA. Telerobotic 3D articulated arm-assisted surgery tools with augmented reality for surgery training. In: Heston TF, editor. Telehealth. IntechOpen: Intech; 2018

[21] Basori AH, AlJahdali HMA. Performance driven-biped control for animated human model with motion synthesis data. Journal of Information Systems Engineering and Business Intelligence. 2018;**4**(2):162-168

[22] Basori AH, Almagrabi A. Towards racing gamification with natural interface for post stroke rehabilitation. Computer Engineering and Applications Journal. 2017;**6**:21-28

[23] Basori AH, Almagrabi AO. Real time physical force-driven hair animation. International Journal of Computer Science and Network Security. 2018;**18**(10):71-77

[24] Basori AH, Al-Sharif AMA-G, AOF AL-O, Almagrabi A. Ba-Rukab OJJoISE, Intelligence B. Intelligence Context Aware Mobile Navigation using Augmented Reality Technology. 2018;**4**:65-72

[25] Basori AH, Dato' b, Abdul Kadir MR, Ali RM, Mohamed F, Kadiman S. Kinect-based gesture recognition in volumetric visualisation of heart from cardiac magnetic resonance (CMR) imaging. In: Ma M, Jain LC, Anderson P, editors. Virtual, Augmented Reality and Serious Games for Healthcare 1. Berlin, Heidelberg: Springer Berlin Heidelberg; 2014. pp. 79-92

[26] Basori AH, Qasim AZ. Extreme expression of sweating in 3D virtual human. Computers in Human Behavior. 2014;**35**:307-314

[27] Basori AH, Tenriawaru A, Mansur ABF. Intelligent avatar on E-learning using facial expression and haptic. Telkomnika. 2011;**9**(1):115-124

[28] Basori AH, Malebary SJ. iDriveAR: In-vehicle driver awareness and drowsiness framework based on facial tracking and augmented reality. In: Gupta N, Prakash A, Tripathi R, editors. Internet of Vehicles and its Applications in Autonomous Driving. Cham: Springer International Publishing; 2021. pp. 93-103

[29] Basori AH, Malebary SJ. Deep reinforcement learning for adaptive cyber defense and Attacker's pattern identification. In: Shandilya SK, Wagner N, Nagar AK, editors. Advances in Cyber Security Analytics and Decision Systems. Cham: Springer International Publishing; 2020. pp. 15-25

[30] Basori AH, Malebary SJ, Firdausiah Mansur AB, Tenriawaru A, Yusof N, Yunianta A, et al. Intelligent socio-emotional control of pedestrian crowd behaviour inside smart city. Procedia Computer Science. 2021;**182**:80-88

[31] Malebary SJ, Basori AH, Alkayal ES. Reinforcement learning for pedestrian evacuation simulation and optimization during pandemic and panic situation. Journal of Physics: Conference Series. 2021;**1817**(1):012008

[32] Mohd Rahim MS, Fata AZA, Basori AH, Rosman AS, Nizar TJ, Yusof FWM. Development of 3D tawaf simulation for hajj training application using virtual environment. In: Visual Informatics: Sustaining Research and Innovations. Berlin, Heidelberg: Springer Berlin Heidelberg; 2011

[33] Tenriawaru A, Mansur ABF, Basori AH, Al-Qurashi Q, Al-Muhaimeed A, Al-Hazmi M. Social awareness and safety assistance of COVID-19 based on DLN face mask detection and AR Distancing. International Journal of Artificial Intelligence Research. 2021;**5**(2):111-122

[34] Manikandan P, Ramesh G, Likith G, Sreekanth D, Prasad GD. Smart nursing robot for COVID-19 patients. In: 2021 International Conference on Advance Computing and Innovative Technologies in Engineering (ICACITE). Greater Noida, India: IEEE; 2021

[35] Murphy R. How robots are helping with COVID-19 and how they can do more in the future. In: 2020 IEEE International Symposium on Safety, Security, and Rescue Robotics (SSRR). Abu Dhabi, United Arab Emirates: IEEE; 2020

[36] Ravuri P, Yenikapati T, et al. Design and simulation of medical assistance robot for combating COVID-19. In: 2021 6th International Conference on Communication and Electronics Systems (ICCES). Coimbatre, India: IEEE; 2021

[37] Riduwan M, Basori AH, Mohamed F. Finger-based gestural interaction for exploration of 3D heart visualization. Procedia—Social and Behavioral Sciences. 2013;**97**:684-690

[38] Sayed AS, Ammar HH, Shalaby R. Centralized multi-agent mobile robots SLAM and navigation for COVID-19 field hospitals. In: 2020 2nd Novel Intelligent and Leading Emerging Sciences Conference (NILES). Abu Dhabi, United Arab Emirates: IEEE; 2020

[39] Siriwardhana Y, Gür G, Ylianttila M, Liyanage M. The role of 5G for digital healthcare against COVID-19 pandemic: Opportunities and challenges. ICT Express. IEEE. Coimbatre, India 2021;**7**(2):244-252

[40] Yusoff YA, Basori AH, Mohamed F. Interactive hand and arm gesture control for 2D medical image and 3D volumetric medical visualization. Procedia—Social and Behavioral Sciences. 2013;**97**:723-729

[41] Zedda A, Gusai E, Caruso M, Bertuletti S, Baldazzi G, Spanu S, et al. DoMoMEA: A home-based telerehabilitation system for stroke patients. In: 2020 42nd Annual International Conference of the IEEE Engineering in Medicine & Biology Society (EMBC). New York: IEEE; 2020

Chapter 3

Perspective Chapter: Evolution of User Interface and User Experience in Mobile Augmented and Virtual Reality Applications

Peter J. Van de Broek, Clement Onime, James O. Uhomobhi and Mattia Santachiara

Abstract

An end-use's experience of any software is typically influenced by the interface presented by the application to the user. For Mixed Reality Environments such as Augmented and Virtual Reality, the user interface is highly visual, and a poor interface can significantly degrade the user experience. Adequate attention is required when designing or creating interfaces and user experience within Mixed Reality Environments as traditional interface design goals and specifications often need to be adjusted. Furthermore, for mixed reality environments on Mobile devices there are additional interface constraints and considerations that would considerably improve the user experience when properly addressed. This research paper discusses the evolution(s) of user interface(s) and user experience of Augmented and Virtual Reality applications on Mobile devices and contributes a framework for improving user interfaces and experience when using Mixed Reality Environments.

Keywords: augmented reality (AR), augmented virtuality (AV), human computer interaction (HCI), mixed reality (MR), user interface (UI), user experience (UX), user experienced interface (UXI), virtual reality (VR)

1. Introduction

User Experience [UX] and User Interface [UI] are major components of any modern software application where interaction with a human is required. While the term UI historically referred to the basic elements that provide input-output functionality [1], such as keyboards, line-printers and visual display units, today, it has broadened to also encompasses elements of visual design including layouts, kinds of prompts/dialog-boxes, fonts/language of text as well as the use of colours and images. There are already voice activated smart or intelligent systems and platforms where the user interface is completely based on audible speech or sound for both input and output.

The UX has been viewed as distinct from the UI and refers to the perception by end-users of the attractiveness and suitability of the software for its intended purpose. In most cases, measurable UX is based on an overall aggregation of both non-abstract and abstract quantities that also include the UI, software functionality and even its response speed. Today, the UX is of paramount importance for any software or (tool) that require user interactions.

Mixed Reality (MR) environments combine both real and virtual (e.g computer generated) objects for presentation within single displays. MR environments have been classified based on the ratio of real to virtual objects within it. Completely virtual reality (VR) environments exists at one end of the continuum while completely real, or physical environments are at the opposite end. In between both ends, the continuum defines arbitrary combinations of both real and virtual objects. When there are more virtual elements than real ones, the environment is classified as Augmented Virtuality (AV) while when there are more real objects than virtual ones, it is known as Augmented Reality (AR) [2].

The AR, AV and VR mixed reality environments can be implemented or displayed using a wide variety of hardware devices that include projectors, visual display units or monitors as well as large wall-sized displays or specialized head mounted displays [3]. This work, however focuses on the display and use of these MR environments on mobile devices, where mobile devices are limited to portable consumer grade ICT devices such as smart-phones/tablets and their associated peripherals such as head or chest mounting units, glasses and watches. That is, we focus on commodity mobile devices such as smart-phones and tablets not including portable or custom (expensive) hardware and equipment [4].

The evolution of UI and UX in MR environments has been heavily influenced by available technology. For example, the MR applications in the 1960s were limited to using wire-frame displays [5]. UX is of particular interest to MR environments as they can easily combine the advantages of both virtual-environments and seamless collaboration [6].

This rest of this chapter provides some background literature review pertinent to the evolution of UI and UX in mobile MR environments, UI/UX frameworks, the unique challenges of the mobile MR environments.

2. Background

Figure 1 presents a redrawing of the Reality Virtuality (RV) continuum first proposed by Milgram and Kishino.

This RV continuum was formulated on a 3-dimensional (3D) taxonomy that incorporated the Extent of real-World Knowledge (EWK), Reproduction Fidelity (RF) and the Extent of Presence Metaphor (EPM) [7]. All of which are fundamental to both UI and UX, that is, increasing EWK translates to a better ability to modelling the real-world (which leads to better UI). Similarly, with increasing RF, real and virtual content becomes more and more indistinguishable (which could to a better UX), and with increasing EPM users' interactions become more natural or better aligned with real environments (which suggests better UX) [8].

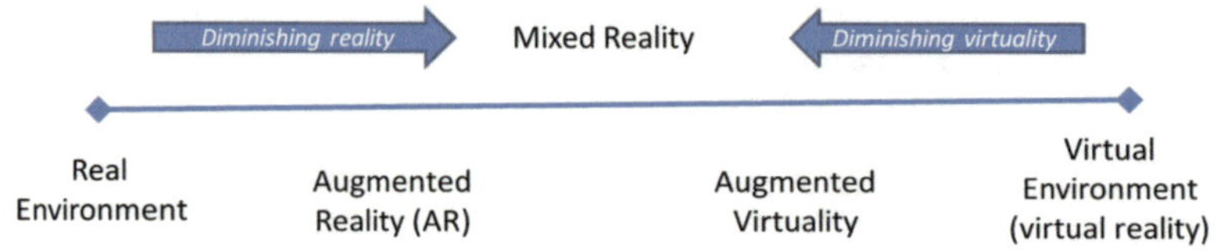

Figure 1.
Redrawn RV contimuum.

Skarbez et al argue that this RV continuum is limited as it describes content only in relation to realism and therefore lacks coherence in the end users' experience or UX [8]. They state that the "mediating" technology, content conveyed, and resulting impact must be considered together to adequately describe MR experiences". Equally pertinent is that the RV continuum was formulated explicitly on visual experiences and visual hardware. Due to rapid advances in hardware and software MR environments are no longer confined to just visually synthesized displays alone but now include experiences that facilitate not only haptic and auditory experiences, with at least exploratory iterations in computer-generated stimuli for all the exteroceptive senses, and it is through interactions with the 5 exteroceptive senses (sight, sound, touch, smell and taste) that users experience MR environments [8].

Despite this limitation, the RV continuum remains a relevant framework for MR research and development today. Indeed for modern applications, there is a need to evolve from more passive or traditional modes of HCI to UI that facilitate multi-sensorial modalities allowing for interactions in virtual worlds, where an interaction modality can be defined as a tangible communication mode [9]. Computer UI aim to enhance interactions with computing systems through various interfaces. Historically UI have evolved from batch interface (punched cards) to command-line user interface, graphical user interface (GUI), web-based user interface (WUI), a subclass of GUI, and recently to touch screens that accept inputs at the touch of a stylus or finger [10], Further evolutions in UI can be classified under the broad category of Post-WIMP UI [11–13], or next generation user interfaces [2]. Such UI employ a variety of novel interaction devices and techniques targeting multi-platform and multi-modal UI that have evolved to address user interactions and experiences in 3-D MR environments, including VR and AR [14], with a need for greater responsiveness, immediacy in feedback and realism within immersive 3-D environments. Examples of NGUI include tangible user interface (TUI), organic user interface OUI, reality-based interface (RBI) and smart material interface (SMI) [15].

2.1 User interface

Traditional UI were predicated on the narrow scope of usability, where cognitive load was reduced, as opposed to users' overall experiences [16]. An early paradigmatic model, that is still ubiquitous, is the window, icon, menu, pointer (WIMP) GUI, facilitated by the introduction of the point and click mouse. The WIMP GUI model developed in the 1980s using interfaces from the computer-as-tool paradigm where a 2-dimensional workspace is presented with direct manipulation of objects in a serial nature [17]. Although the WIMP GUI was adapted and popularised by Macintosh in the 1980's it is still the most dominant type of GUI in modern desktop computers [11]. Reasons for the ubiquity of this GUI include its' effectiveness in facilitating common office tasks [11]. Other advantages are its ease of use due in part to exploitation of muscle memory and image recognition and commonality across applications with widespread accessibility for a range of users, facilitating the creation of a de facto standard [12]. With the introduction of WIMP interactions with computing hit the mainstream. Before this time van Dam [13] argues that there were two previous generations of user interfaces, placing WIMP in the third generation of UI. UI at this time were optimised to the available hardware, although it is argued that the first generation in the 1950s and 1960s were not UI in the strict sense as there was no interaction with users per se as computers were used in batch mode with punched-card inputs and line-printer output. Between the 1960s and 1980s van Dam [13] highlights the evolution of the second generation of UI, in which for the first-time users could interact with computing systems by typing in parameter defined commands on mechanical alphanumeric displays using timesharing on mainframes and microcomputers. Such systems were founded on operating systems such as DOS and UNIX, with command line shells and device drivers. In the DOS OS the device driver has responsibility for input/output operations, and uses blocks, with their own address, to store information [18] in disks. In this way the user controls all system software through the DOS UI that allows for graphical displays on the monitor. DOS was a forerunner to the GUI and is a command line interface (CLI) system. Key considerations of earlier iterations of UI were responsiveness and immediate feedback to user inputs, increasing functionality through human computer interactions [HCI], where functionality was the key paradigm However WIMP UI have several limitations. As the complexity increases, with additional icons and widgets added, the UI becomes more cumbersome and harder to use, with the serialised nature of the interface separating the user from the perceptions of real time working [11] and preventing parallel inputting [19]. In addition, the UI is predicated on a 2-D paradigm, with 2-D input devices and desktop metaphor and do not innately transpose into 3-D environments [11]. Such limitations have become more pronounced. Evolutions in processing and graphical processors, leading to advancements in software and hardware and iterations in designing and development of more appropriate UI have taken place, with developments in gaming having a major input. With the increasingly widespread proliferation of gaming – from handheld to desktop and online collaborative platforms utilising immersive 3-D worlds, HCI had to evolve in which the overall concept of UX became more of a consideration. As Bonnardel [20] argues as UI evolve in response to e.g., games and 3-D environments novel techniques must be used that are future focused. Equally as importantly such UI need to go beyond functionality and elicit feelings of fun and enjoyments for users [21], highlighting the significance of the overall UX, which is enhanced through increased emotional investment, or affective perceptions [22–24]. As Tractinsky et al. [23] report correlations exist between users' perceptions of the aesthetics of the HCI system and its usability. Jakubowski [25] sums this up when stating that the most important aspect of HCI is the influence of a good UX experience on the user productivity. McCarthy and Wright [26] define UX as a qualitative experience while interacting with products. The logical argument being that as users' qualitative experiences increase, through more immersive, multi-modal and realistic UI, HCI improve, whether they be purely functional, for enjoyment or for educational purposes. Early gaming experience, such as Pong and Space Invaders came to the forefront in the 1970s. As Sahay et al. [27] report, although these early gaming iterations, like all games have the ability to engage people, due to a lack of processing power for example, they lacked features, such as shading, texture, realism and dimensionality, with unattractive and unrealistic graphics. With improvements in software and hardware not only has gaming made huge strides with graphics becoming more realistic, but modern gaming also now incorporates artificial intelligence (AI), Evolutionary advances in portability, range of consoles, including mobile, and network-based gaming [27] has culminated in modern online games, with more responsive controllers that take place in virtual environments with the ability to compete against remote opponents [28]. This in turn has increased the appeal of gaming through immersion. As Jennett et al [28] argue not only does immersion transcend the idea of flow, cognitive absorption [CA] and presence, it is a measure of engagement, engrossment and total immersion, as also reported by Brown and Cairns [29]. Csikszentmihalyi [30] argues that flow happens when individuals are completely engrossed in an activity to the detriment of other things. Thus, the concept of immersion involves losing track of time and cognisance of the real world, involvement and becoming lost in the game, or virtual environment, and is dependent on a good gaming experiences [28]. Thus, the overall UX is enhanced, mediated through more intuitive, interactive, realistic, multi-modal and responsive UI. As Brown and Cairns [29] state "engagement, and therefore enjoyment through immersion, is not possible if there are usability and control problems. Essentially there needs to be an invisibility of the controls for total immersion to take place." In other words, for enhanced UX, UI need to evolve to become unobtrusive, intuitive to use and multi-sensorial, so that UI are subsumed within the interactive experience. Such advances in UI and increased UX are also apposite to interactions with MR environments.

2.2 Considerations for mixed reality environments

Although most applications still try to cope with a WIMP-style user interface and two-dimensional input, devices with multiple degrees of freedom are still rare [11]. However with the growth in 3-D applications and MR environments UI are evolving to meet the needs of users interacting with such environments. The overriding difference is that in MR environments the UI has to shift away from virtual interfaces designed to mediate interactions with computer systems to interfaces that combine both real and virtual environments and objects, dispersed at any point along the MR continuum. The ultimate aim is seamless interaction in the same environment. In this way UI in MR environments need to be able to integrate with a real environment where static and dynamic information streams are combined at runtime [2]. Billinghurst et al. [31] sum this up when stating that "AR interfaces are designed to enhance interactions in the real world." UI designed to work within 3 dimensions contain greater complexity and need to be multi-modal and sensorial in nature. They require more degrees of freedom (DOF) and greater user efficiency due to the greater number of non-serial tasks, involving parallelism [11]. Additionally, due to the wide range of MR environments and possible applications more interactions between users and the environments are needed and a wider range of UI are needed. As Bowman et al. [32] state performance of UI in such environments is task and environment dependent with specific UI, targeted at displays that may be fully immersive or semi-immersive, being needed. Such UI are dependent on ergonomics and the target device with input/ output interactions in MR interfaces trending towards increasing naturalness becoming more intuitive and seamless. Complexities in UI applicable to MR and 3-D environments are due to several factors. These include the range of applicable input devices, which may be discrete, continuous or a hybrid of both, alongside the navigational options potentially available, ranging from more general exploration of such environments to searching for specific locations as well as more precise manoeuvring [32]. In addition, interfaces in such environments need to allow for the ability to interact with, and manipulate, objects in such environments. This can involve zooming and rotation with direct user control, physical control and/or virtual control [33]. One central feature of MR UI is the integration with a real environment. The application requires information about objects and spaces, whose geometry and behavior is not under the control of the designer but must be acquired from the real environment. Real objects can be subject to real-world manipulation [e.g., in a maintenance task] or external forces. Therefore, it must be possible to track state changes in the environment. In practice the "real world" model of a mixed reality application often consists of a combination of static information [e.g., geometry of the environment that is assumed to be fixed] and dynamic information [e.g. position and orientation information for the user and central objects] that is acquired by sensors at runtime. Sherman and Craig [33] describe direct user control as mimicking real world interaction, physical control that uses real devices and virtual control using virtual devices [11]. All these factors point to the fact that UI applicable to MR environments, unlike WIMP interfaces, need high bandwidth as well as efficient processors. In addition, continuous sampling and processing, probabilistic decoding and recognition that can unify input from parallel channels through multi-modal interfaces are needed [12]. As Van Dam [12] describes UI in MR environments need to facilitate body part tracking, gesture and speech recognition as well as haptic force input and feedback devices. Sub-subsections can also be used throughout the manuscript.

2.3 Considerations for mobile devices

Due to the small screen size, lack of memory, low to moderate processing power, smaller and fewer buttons and limited battery power, alongside the array of sensors, UI for mobile devices have numerous constraints, which can affect overall levels of UX. Subramanya and Li [33] classify these types of constraints as device related constraints with user related constraints including limited attention spans affected by mobility, change in locations and contexts and users' idiosyncrasies. Chong et al. [34] argue the UI and mobile device size are one of the most significant factors in mobile device design and report on the use of a single-layer touch screen UI as opposed to the more conventional multi-layer UI, with promising results in increasing overall UX. The use of low-level computer languages, termed code optimization [34] also helps in reducing strains on available memory, as does the use of touchscreen UI, as opposed to mouse based and command-based UI topologies. This use of low-level language and single-layer UI can potentially overcome issues due to the noted complexities involved in developing applications and UI across various mobile platforms. Such mobile platforms can be incompatible, alongside the variety of programming languages and hardware differences as reported in [35]. Touchscreen UI obviate the need for physical keyboards, thereby maximising available screen sizes whilst at the same time increasing mobility with concomitant reductions in device sizes, as argued in [34]. Touchscreen UI are also aesthetically more pleasing and intuitive to use, thereby potentially facilitating increased UX. As Dunlop and Brewer [36] report with the increasing proliferation and popularity of mobile devices issues of widening access to powerful computing services and resources through the UI need to be overcome when designing UI with good UX. In addition, alongside the small visual displays mobile devices have had poor interaction facilities, including audio and limited input/output (I/O) [36] which create challenges posed by mobile device UI, which are exacerbated by network access issues. However, with advances in mobile device software and hardware leading to increased performance, effective UI designs have and are being proposed and developed. As Choi [37] report such UI can be classified into hardware and vision based, with vision-based UI receiving more focus due to not needing extra technical equipment or physical sensors. Such extra equipment may be inconvenient and relatively inaccurate [37], potentially leading to less well perceived UX due to the need to interact with additional layers, increasing the complexity of the HCI.

3. Evolution

Skarbez et al. [8] have suggested or proposed a revised RV continumm as shown in **Figure 2**, based on the idea that MR environments do not affect the interoceptive senses, can be termed as "external" MR environments. It is only when technology can also stimulate internal senses that virtual environments can be separate from the MR continuum, This revised continuum introduces a discontinuity within VR environments. That is, External Virtual Reality (EVR) environments, next to AV

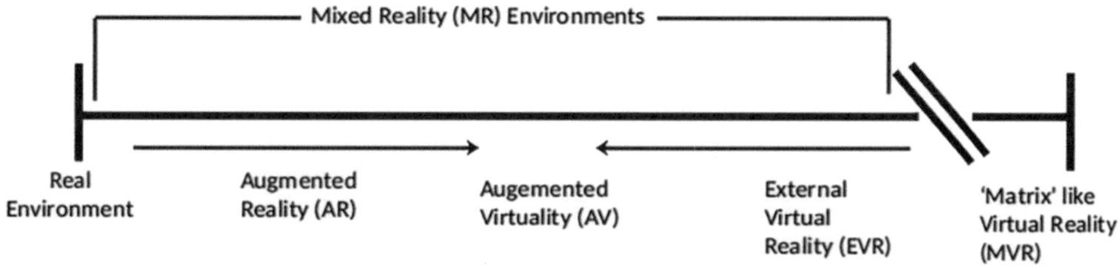

Figure 2.
Revised RV contimuum.

environments and the discontinuity before a 'Matrix-like' VR environment at the extremity. This allows the continuum to take cognisance of VR environments that focus on stimulations of the interoceptive senses, while external virtual reality environments are remain MR environments. Note, in the revised RV continuum, the EVR is equivalent to the "Virtual Environment" extremity in **Figure 1** and is still part of the MR environment.

Skarbez et al consider any form of technology-mediated reality as MR [8]. This is comprehensible when mediated reality encompasses users' interactions with the world around them, through the use of technology as an extension of users' minds and bodies [38]. Such arguments expand on MR experiences as going beyond just visual interactions [7]. This is implicit when Paradiso and Landay [38] define extended reality (XR) as a MR environment that involves the union between sensor/actuator networks and shared online virtual worlds. To take account of the interaction between sensor networks and virtual worlds and how a user experiences them, Skarbez et al have re-defined MR as an environment "in which real world and virtual world objects and stimuli are presented together within a single percept," where different senses, not just sight, may be affected.

As van Dam [12] argues evolutions in UI need to match human perceptual, cognitive, manipulative and social abilities. At the same time interactions need to be as seamless and natural as possible, thereby increasing overall UX. What these evolutionary advancements perhaps highlight, concomitant with advancements in gaming and 3-D MR environments, is that an overall paradigm shift has, and is occurring, from UI that can be viewed as purely functional and non-interactive, such as line printers and earlier iterations of Visual Display Units (VDU) to Graphics Processing Units (GPU) and onto more encompassing, immersive and interactive UI, in which UX is of greater importance. In simple terms non-interactive and more functional UI, that may not have such a high degree of UX, involve using and displaying texts and images as labels, that provide contextual information about objects, images etc., In contract interactive UI, where UX is more of an important paradigm, include buttons, toggles, sliders and other components that facilitate interactions with UI tools, such as icons on touch screens. Evolutionary UI inputs have been aimed at increasing productivity and efficiency through increased interaction, starting with the mouse developed by Douglas Engelbart in 1964. The mouse allowed for greater computer screen interaction in 2-D worlds. Since then, to input UI devices have evolved from the mouse, in line with advancements in gaming and 3-D MR environments, in which improvements in UX are paramount, to include game controllers, motion controllers, hand tracking devices to the Litho controller in 2018 [39]. According to Hillman [39] the Litho controller is an innovative solution that may be able to address shortcomings in hand input or traditional controllers helping with hand fatigue and increasing haptic feedback.

The beginnings of HCI in the late 1950s and early 1960s involved "batch processing" in which programs and data were read from cards or tape (paper or magnetic) until termination with a printed output, via line printers. VDUs superseded such operations, which were still restricted to scrolling commands and responses one line at a time [40]. Research carried out by Ivan Sutherland in the early 1960s led to the development of more powerful computing systems and graphics, with developments in GPU, or graphics cards, as well as developments and evolution in object-oriented programming concepts [41]. Object oriented programming is the fundamental paradigm in the C# computer programming language and focuses on data objects instead of functions and logic, as well as providing the inspiration for the development of Object-Oriented User Experience (OOUX), that classifies objects that users interact with first, before assigning actions to such objects [41]. This seems particularly apposite for MR environments and UI. As Hillman [41] argues OOUX allows for better interaction with spatial 3-D worlds. GPU developments have facilitated accelerations in graphical rendering with many pieces of data being processed simultaneously, leading to more powerful and faster computers. Such evolutionary developments led to the creation of the Xerox Alto in 1973. Although too expensive for widespread use, the Alto supported the use of GUI, as opposed to prototypes [42], as well as being the precursor for evolutionary advancements in gaming and ultimately developments in 3-D immersive MR environments. The emergence of GUI was seen as a disruptive revolution in HCI, being more advantageous and attractive in the early iterations to new users. It was not until 1985, with the release of the Apple Mac, that GUI started to be seen as being successful, and even more importantly with the successful release of Windows 3.0 in 1990 were GUI more widely accepted by government agencies and businesses who controlled research funding [40]. Also, during the late 1960s the first computer aided design [CAD] systems were promulgated with the development of 2-D and 3-D wireframe graphics, with all CAD systems now being based on a windows – menu interface with 3-D models [43]. Wireframe graphics map models, images and objects in 3-D, comprising vertices and edges [44] using triaxial [x,y,z] cartesian coordinates, where the z coordinate represents the height. Wireframe graphics allow for simplicity in presentation and flexibility in the use of colour [45]. Vertices are a collection of the 3-D coordinates connected together into triangles which can contain information such as colours, textures and directions [46], which are displayed through rendering and shading. Evolutions and developments in rendering and shading have further enhanced graphics and GUI. Rendering is the process of generating images and shaders are programmes that take meshes and textures etc. as inputs to generate the outputted image [47]. **Figure 3** illustrates how rendering works in Unity.

4. Framework

Numerous frameworks have been proposed for designing and developing UI taking into account UX. Indeed, with the advent of immersive 3-D MR environments and their concomitant UI, in which UX is an increasingly more important concept, UX has in many cases subsumed UI as part of the design process and framework. In this way a more holistic approach can be taken in which UI and UX are combined into one paradigm, which for the sake of this paper can be termed user experienced interface [UXI]. As Hassenzahl and Tractinsky [48] argue, overall UX is influenced by end users' internal states, including predispositions, expectations, needs, motivations and emotions; as well as the characteristics of the designed interface, including complexity, purpose, usability and functionality, and; the context within which the interaction occurs, be it the organisational or social setting and meaningfulness of the activity.

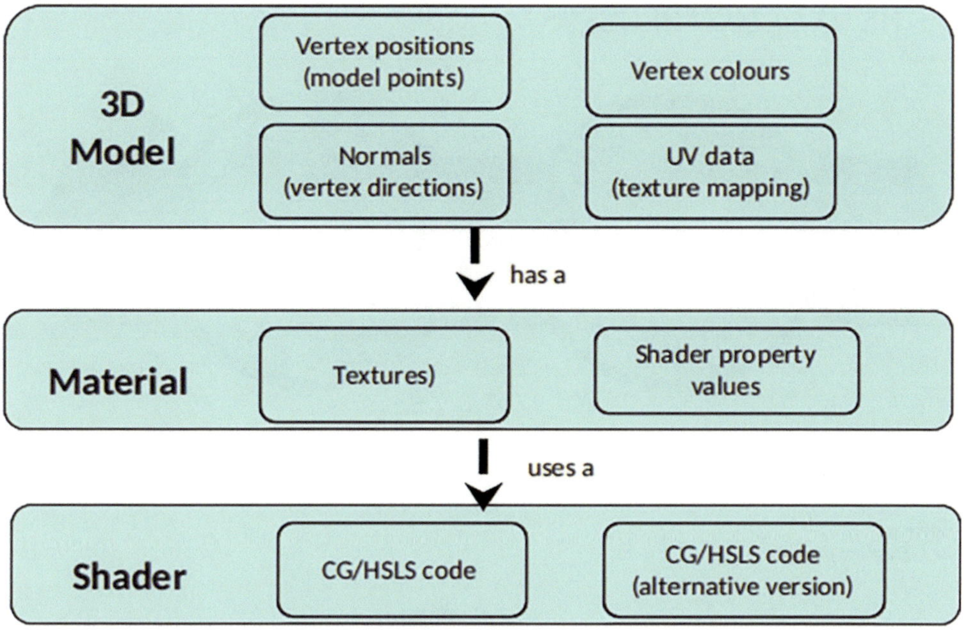

Figure 3.
The rendering workflow.

The design of the interface, or system, can be conflated with UI, whereas the context in which the interaction occurs can be conflated with immersive experiences in 3-D MR environments, whether that involves mobile devices or not, all encompassed within the overarching UX paradigm. Going a step further Hassenzahl [49] put forward a model for UX design in which users perceive interactions with products in two dimensions: hedonism and pragmatism. The hedonic aspect refers to the users' interactive experiences and the ability of the system to support what has been termed "be-goals, which correlate more to the enjoyability, and emotions involved with interaction. In contrast the pragmatic aspect refers to the perceived ability of users' interactions with the system to support "do-goals," which correlate more to functionality and efficiency domains.

Hillman [39] argues in favour of the importance of frameworks for designing and deploying UX, in regard to MR, or XR, environments and applications. Such frameworks can be enhanced by incorporating integrated development environments [IDE] with built in presets that allow for faster prototyping and iterations. One such appropriate IDE is the Unity 3-D game engine which provides opportunities for developing UX for MR environments. Unity is a software framework, that provides a set of tools for "developers around the world to create rich, interactive, 2-D, 3-D, VR and AR experiences" (Unity public relations fact page, n.d.), negating the need for the construction of virtual spaces from the ground up [50]. Examples of preset built in core functionality includes the AR Foundation package which provides presets and plugins to enable development of immersive AR applications, to mobile devices [both Android and Apple], as well as web based and wearables. This is especially true when having to deal with mobile device considerations, which have been outlined in Section 2.3.

Other important considerations in UX design include the interaction between user needs, whether that is private enterprise or public organisations, and business goals along with the fundamentals, of end users wants and needs, ideation, prototyping, testing and implementation, with iteration [39]. **Figure 4** illustrates this principle.

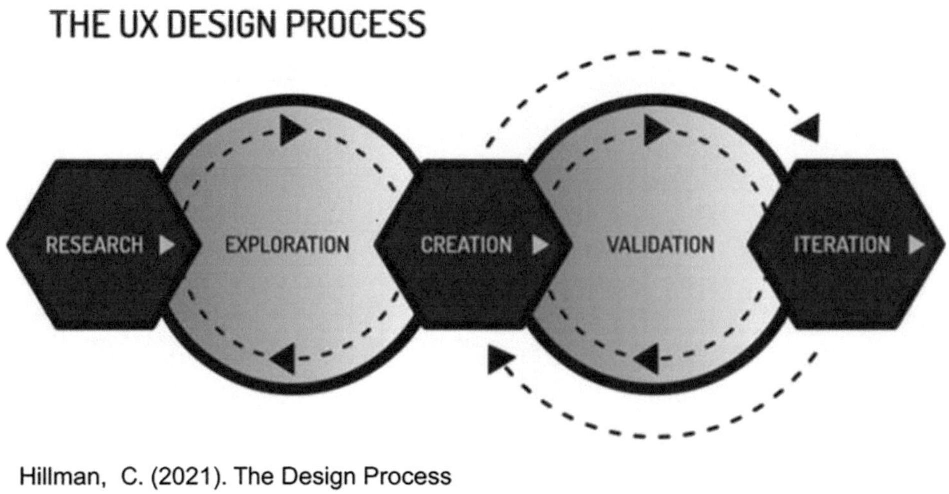

Figure 4.
The UX design process.

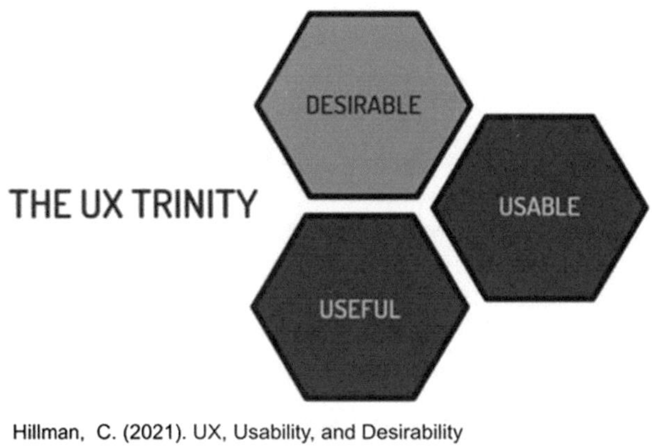

Figure 5.
The Trinity of UX.

By evaluating end users' needs and wants the useability, usefulness and desirability of the UX in MR can be determine. Hillman [39] classifies this as the trinity of UX design and is illustrated in **Figure 5**.

To arrive at this trinity and design UX effectively and efficiently it would therefore seem that collaboration with end users is another key facet of any framework, much like the incorporation of an IDE, such as Unity. By doing this a modified UXI framework built on foundations outlined by Hillman's UX design proces is proposed. This is illustrated in **Figure 6**.

In this framework aimed at MR applications, UI and UX are merged into one holistic paradigm: UXI. In the consultation phase a collaborative approach is needed to ascertain end users' needs and wants, be the desired goal an AR education app or a VR app aimed at private enterprise, alongside usability, more specifically ease of use. This leads to the ideation phase in which the development of MR application will take place using the Unity IDE, with built in packages aimed at the development of MR immersive environments. One such example is the AR Foundations package, which contains monobehaviours for such things as planar surfaceÂ detection; point clouds; reference points: arbitrary positions and orientations that devices track; light estimation: estimates for average colour temperature and brightness in

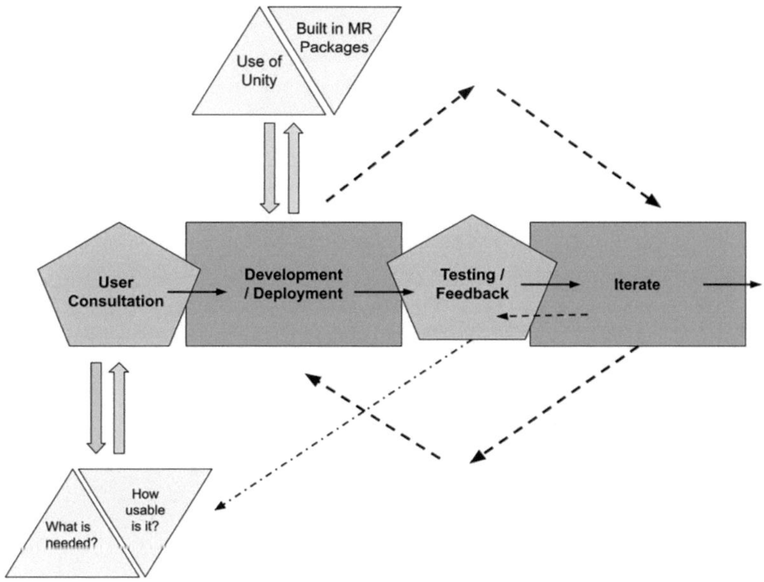

Figure 6.
UXI framework.

physical spaces, and world tracking: tracking the device's position and orientation in physical spaces [51]. Post development, deployment of the MR UXI will take place, after which feedback will be a key factor leading to iteration and continual deployment, with or without changes.

5. Conclusions

The advances in computing and HCI have led to changes in how humans use and interact with computers and other devices, through evolutionary developments in UI and UX. Latterly mobile devices, such as tablets and especially smart phones have become widespread in their proliferation and use. Such mobile devices now have many of the capabilities of larger computers and laptops, albeit with limitations, such as lack of memory, power and smaller screen sizes. Due to advancements in mobile devices such evolutions in UI and UX have been even more pronounced, leading to the blurring between UI and UX, which can no longer be viewed as discrete and separate. UI have been subsumed into overall concepts of UX and frameworks for development and deployment. This is even more pertinent with the advent of MR 3-D immersive environments and how users interact with them. This occurs in a variety of contexts, with interactions occurring more and more on mobile devices. This paper discussed evolutions in UI and UX and how they merged into UXI and proposed a framework for the development of MR UXI, applicable to mobile devices, as well as other devices in general. The next steps are to use this framework in the development of a MR environment UXI for mobile devices and analyse the results.

Author details

Peter J. Van de Broek[1], Clement Onime[2]*, James O. Uhomobhi[3,4] and Mattia Santachiara[5]

1 Artificial Intelligence [AI] Research Centre, Ulster University, Northern Ireland, UK

2 Abdus Salam International Centre for Theoretical Physics, Trieste, Italy

3 School of Engineering, Ulster University, Northern Ireland, UK

4 School of Computing, Ulster University, Northern Ireland, UK

5 Santasco, Via Dardano Fenili, Reggio Emilia, Italy

*Address all correspondence to: onime@ictp.it

References

[1] Bolton W. Programmable Logic Controllers. 6th ed. USA: Newnes; 2015

[2] Stöcklein J, Geiger C, Paelke V, Pogscheba P. A Design Method for Next Generation User Interfaces Inspired by the Mixed Reality Continuum. Berlin, Heidelberg: Springer Berlin Heidelberg; 2009. pp. 244-253

[3] Onime C, Uhomobhi J, Bornschlegl MX, Engel FC, Bond R, Hemmje ML. Cost Effective Visualization of Research Data for Cognitive Development Using Mobile Augmented Reality. Cham: Springer International Publishing; 2016. pp. 35-49

[4] FitzGerald E, Adams A, Ferguson R, Gaved M, Mor Y, Thomas R. Augmented reality and mobile learning: The state of the art. In: Specht M, Sharples M, Multisilta J, editors. 11th World Conference on Mobile and Contextual Learning (mLearn 2012), Helsinki; 2012. pp. 62–69

[5] Onime C, Uhomoibhi J, Wang H, Santachiara M. A reclassification of markers for mixed reality environments. The International Journal of Information and Learning Technology. 2021;**38**(1):161-173

[6] Billinghurst M, Kato. Collaborative Mixed Reality. In: Proceedings of the International Symposium on Mixed Reality. Yokohama, Japan; 1999. pp. 261–284

[7] Milgram P, Kishino. A taxonomy of mixed reality visual displays. IEICE Trans Information and Systems. 1994; **12**:1321-1329

[8] Skarbez R, Smith M, Whitton M. Revisiting Milgram and Kishino's Reality-Virtuality Continuum. Frontiers in Virtual Reality. 2021;**2**:27

[9] Appert D. Conception et évaluation de techniques d'interaction non visuelle optimisées pour de la transmission d'information [Theses]. Université Paul Sabatier - Toulouse III; 2016. Available from: https://tel.archives-ouvertes.fr/tel-013 94411

[10] Barnes S. User friendly: A short history of the graphical user interface. Sacred Heart University Review. 2010; **16**(1):4. Available from: http://digitalcommons.sacredheart.edu/shureview/vol16/iss1/4

[11] Biström J, Cogliati A, Rouhiainen K. Post-wimp user interface model for 3-D web applications; 2005. Available from: https://www.semanticscholar.org/paper/Post-WIMP-User-Interface-Model-for-3-D-Web-Bistr%C3%B6m-Rouhiainen/59b33d4cf54a11156a091ba98b003dfe900c98cc

[12] Van Dam A. Beyond WIMP. IEEE Computer Graphics and Applications. 2000;**20**(1):50-51

[13] Van Dam A. Post-WIMP user interfaces. Communications of the ACM. 1997;**40**(2):63-67

[14] Shaer O, Jacob R, Green M, Luyten K. User interface description languages for next generation user interfaces. In: Proceeding of the Twenty-Sixth Annual CHI Conference Extended Abstracts on Human Factors in Computing Systems—CHI'08. 2008. pp. 3949–3952

[15] Georgakopoulou N, Zamplaras D, Kourkoulakou S, Chen CY, Garnier F. Exploring the virtuality continuum frontiers: Multisensory and magical experiences in interactive art. In: Proceedings of the 14th International Joint Conference on Computer Vision, Imaging and Computer Graphics Theory and Applications; 2019. pp. 175–182

[16] Bastien J, Scapin D. A validation of ergonomic criteria for the evaluation of human-computer interfaces. International Journal of Human-Computer Interaction. 1993;**4**(2): 183-196

[17] Beaudouin-Lafon M. Designing interaction, not interfaces. Proceedings of the Working Conference on Advanced Visual Interfaces; 2004. pp. 15–22

[18] Imam A, Rousan T, Aljawarneh S. An expert code generator using rule-based and frames knowledge representation techniques. IEEE. 2014: 1-6

[19] Odell D, Davis R, Smith A, Wright P. Toolglasses, marking menus, and hotkeys: a comparison of one and two-handed command selection techniques. In: GI '04 Proceedings of Graphics Interface 2004. Research Collection School Of Computing and Information Systems; 2004. pp. 17–24. Available from: https://ink.library.smu.edu.sg/sisresearch/815

[20] Bonnardel N. Designing future products: what difficulties do designers encounter and how can their creative process be supported? Work. 2012;**41**(Suppl 1):5296-5303. DOI: 10.3233/wor-2012-0020-5296. PMID: 22317539

[21] Norman D. Emotion and design: Attractive things work better. Interactions Magazine. 2002

[22] Schenkman BN, Jönsson FU. Aesthetics and preferences of web pages. Behaviour & Information Technology. 2000;**19**(5):367–377. DOI: 10.1080/014492900750000063

[23] Tractinsky N, Katz A, Ikar D. What is beautiful is usable. Interacting with Computers. 2000; **13**(2):127-145

[24] Van Der Heijden H. Factors influencing the usage of websites: The case of a generic portal in The Netherlands. Information & Management. 2003;**6**:541-549

[25] Jakubowski M. User experience as a crucial element of future simulation and gaming design. In: Developments in Business Simulation and Experiential Learning: Proceedings of the Annual ABSEL Conference. Vol. 42; 2015.

[26] Mccarthy J, Wright P. Technology as Experience. Interactions. 2004;**11**(5): 42–43

[27] Sahay A, Choudhury T, Kumar V. Hierarchical Evolution of Digital Games. In: 2019 International Conference on Computational Intelligence and Knowledge Economy (ICCIKE); 2019. pp. 381–386

[28] Jennett C, Cox A, Cairns P, Dhoparee S, Epps A, Tijs T, et al. Measuring and defining the experience of immersion in games. International Journal of Human-computer Studies. 2008;**9**:641-661

[29] Brown E, Cairns P. A Grounded Investigation of Game Immersion. In: CHI'04 Extended Abstracts on Human Factors in Computing Systems. CHI EA'04. New York, NY, USA: Association for Computing Machinery; 2004. pp. 1297–1300. DOI: 10.1145/985921.986048

[30] Csikszentmihalyi M. In: Toward a Psychology of Optimal Experience. Dordrecht, Netherlands: Springer; 2014. pp. 209–226. DOI: 10.1007/978-94-017-9088-814

[31] Billinghurst M, Clark A, Lee G. A survey of augmented reality. A survey of augmented reality. Foundations and Trends in Human. Computer Interaction. 2015;**8**(2-3):73-100

[32] Bowman D, Kruijff E, Laviola J, Poupyrev I. An introduction to 3-D user interface design. Presence: Teleoperators and Virtual Environments. 2001;**10**(1):96-108

[33] Sherman W, Craig A. Understanding virtual reality: interface, application and design. Morgan Kaufmann series in computer graphics and geometric modeling. San Francisco, CA: Morgan Kauffman; 2003

[34] Chong P, So P, Shum P, Li X, Goyal D. Design and implementation of user interface for mobile devices. IEEE Transactions on Consumer Electronics. 2004;**50**(4):1156-1161

[35] Madari I, Lengyel L, Levendovszky T. Modeling the user interface of mobile devices with DSLs. In: 8th International Symposium of Hungarian Researchers on Computational Intelligence and Informatics. Budapest, Hungary; 2007. pp. 583–589

[36] Dunlop M, Brewster S. The challenge of mobile devices for human computer interaction. Personal and Ubiquitous Computing. 2002;**6**(4): 235-236

[37] Choi J, Park H, Park J, Park JI. Bare-hand-based augmented reality interface on mobile phone. In: 2011 10th IEEE International Symposium on Mixed and Augmented Reality; 2011. pp. 275–276

[38] Mann S, Havens J, Iorio J, Yuan Y, Furness T. All Reality: Values, taxonomy, and continuum, for Virtual, Augmented, eXtended/ MiXed (x), Mediated (x,y), and Multimediated Reality/Intelligence. In: AWE 2018, Santa Clara, California; 2018. Available from: http://genesis.eecg.toronto.edu/all.pdf

[39] Hillmann C. In: The Rise of UX and How It Drives XR User Adoption. Berkeley, CA: Apress; 2021. pp. 73–116. DOI: 10.1007/978-1-4842-7020-23

[40] Grudin J. From tool to partner: The evolution of human-computer interaction. Synthesis Lectures on Human-Centered Interaction. 2017;**10**:i–183

[41] Hillmann C. In: UX and Experience Design: From Screen to 3D Space. Berkeley, CA: Apress; 2021. pp. 117–155. DOI: 10.1007/978-1-4842-7020-24

[42] Sears A, Jacko J. The Human-Computer Interaction Handbook: Fundamentals, Evolving Technologies and Emerging Applications. CRC Press; 2007

[43] Dai F. Introduction: Beyond Walkthroughs. Berlin, Heidelberg: Springer Berlin Heidelberg; 1998. pp. 1-9

[44] Cates N. A comparative survey of computer graphics applications in mechanical design. Lehigh University; 1982. Available from: https://core.ac.uk/download/pdf/228649677.pdf

[45] Kellie A. The wireframe model-showing 3D structure with open space. In: Proceedings, 66th Midyear Conference, Engineering Design Graphics Division, ASEE; 2010. pp. 1–9

[46] Learn U. Rendering and Shading – Unity Learn; 2021. Available from: https://learn.unity.com/tutorial/rendering-and-shading#5c7f8528edbc2a002053b539

[47] Hocking J. Introduction to Shaders in Unity. 2020. Available from: https://www.raywenderlich.com/5671826-introduction-to-shaders-in-unity

[48] Hassenzahl M, Tractinsky N. User experience: A research agenda. Behaviour & Information Technology. 2006;**2**:91-97

[49] Hassenzahl M. The hedonicpragmatic model of user

experience. Towards a UX manifesto. 2007

[50] Ward J. What is a game engine? GameCareerGuide; 2008. Available from: https://www.gamecareerguide.com/features/529/whatisagame.php

[51] unity3d com D, editor. About AR Foundation: Package Manager UI website. 2018. Available from: https://docs.unity3d.com/Packages/com.unity.xr.arfoundation@1.0/manual/index.html

Section 2

Haptic for Medical Purposes

Chapter 4

A Proposal of Haptic Technology to be Used in Medical Simulation

Pablo Sánchez-Sánchez, José Daniel Castro-Díaz, Alejandro Gutiérrez-Giles and Javier Pliego-Jiménez

Abstract

For medical training aims, tele-operation systems have inspired virtual reality systems. Since force sensors placed on the robotic arms provide interaction force information that is transmitted to the human operator, such force produces a tactile sensation that allows feeling some remote or virtual environment properties. However, in the last two decades, researchers have focused on visually simulating the virtual environments present in a surgical environment. This implies that methods that cannot reproduce some characteristics of virtual surfaces, such as the case of penetrable objects, generate the force response. To solve this problem, we study a virtual reality system with haptic feedback using a tele-operation approach. By defining the operator-manipulated interface as the master robot and the virtual environment as the slave robot, we have, by addressing the virtual environment as a restricted motion problem, the force response. Therefore, we implement a control algorithm, based on a tele-operation system, to feedback the corresponding force to the operator. We achieve this through the design of a virtual environment using the dynamic model of the robot in contact with holonomic and non-holonomic constraints. In addition, according to the medical training simulator, before contact, there is always a free movement stage.

Keywords: Haptics, virtual reality system, bilateral teleoperation, holonomic constraint, non-holonomic constraint, position-force control

1. Introduction

Teleoperation and virtual reality systems are intrinsically related since they make a human operator interact with the environments without being in physical contact with them. In the first one, these environments are real, while in the second one, we generate them in a computer simulation. However, both types of systems must make the operator perceive, as realistically as possible, the characteristics of the remote or virtual environment. Some variables used to reproduce these characteristics are position and force, which provide visual and haptic feedback, respectively. The biggest difference between them is the procedure of generating the information received by the operator. In the teleoperation case, both signals exist physically and are transmitted via a control algorithm. The algorithm receives both signals from sensors. Both signals do not exist in virtual reality and we must generate computationally them.

Stimulating the senses of sight and touch, as precisely as possible, is essential during the interaction process, since they are the principal channels with which the operator perceives the world around them. For teleoperation, in the visual case, the communication comes directly from the environment or, if the operator is in a remote place, using a camera and a computer screen. The virtual reality system generates the environment through a digital simulation, and the operator receives the visual information through a screen. We need additional devices since the tactile issue is more complex. Those devices must be capable of transmitting the generated forces in the environment. Such a process implies including haptic robots in the systems because of their capability to generate forces and torques that the human operator can perceive in a tactile way. We need for teleoperation systems two physical robots, while in a virtual reality system we only need one robot and the virtual environment as the other.

The medical area has actively seized on both teleoperation and virtual reality systems. In the first case, a specialist can perform surgery procedures over long distances, eliminating the need for the physical presence of either the physicians or patients in the same location [1]. Practically, the specialist can examine or operate on the patient at a different geographic location without having to travel.

In **Figure 1**, we showed the emblematic *Da Vinci Surgical Robotic System*. Operating based on a master-slave control concept. The system provides the medical expert with a realistic operating environment that includes a high-quality stereo visualization and a human-machine interface that directly transfers the doctor's hand gestures to the instrument inside the patient [2].

In the second case, someone has widely used virtual reality systems for medical training simulation. With the development of computer graphics, nowadays practically any surgical procedure can be visually simulated. Minimally invasive surgery has been the most beneficial area. It has implemented virtual environments in laparoscopy, neurosurgery, and urology, to name a few [3]. As an example, in **Figure 2** we showed *the simulator of Transurethral Resection of the Prostate TURP Mentor*. Clinicians can improve their skills without endangering their patients to avoid *Live practice with humans* by using this system. Tactile feedback inclusion is an important factor in the improvement of such skills. It must synchronize tactile feedback with both *the virtual reality simulation* and *the operator's movements*.

The challenge for researchers in graphic computing and control systems is to design mathematical tools that fit with the object's physical characteristics to be simulated within the virtual environment. Regarding the visual feedback, the position where such objects are located is essential for the operator to perceive his movements in the Cartesian virtual space. About the force, reproducing rigidity and softness takes special interest when the virtual environment includes penetrable and nonpenetrable objects. Here, the complexity of the mathematical tools increases because their physical laws are not always easy to simulate on a computer. For this reason, it must establish an interchange between visual and haptic realism because of the finite capacity of digital processing.

1.1 State-of-the-art

Teleoperation and virtual reality applications have developed in areas as different as *automotive* and *video games*. However, an important application is for medical surgery [4], where *the operator* on *the master side* needs to be sure of the force he

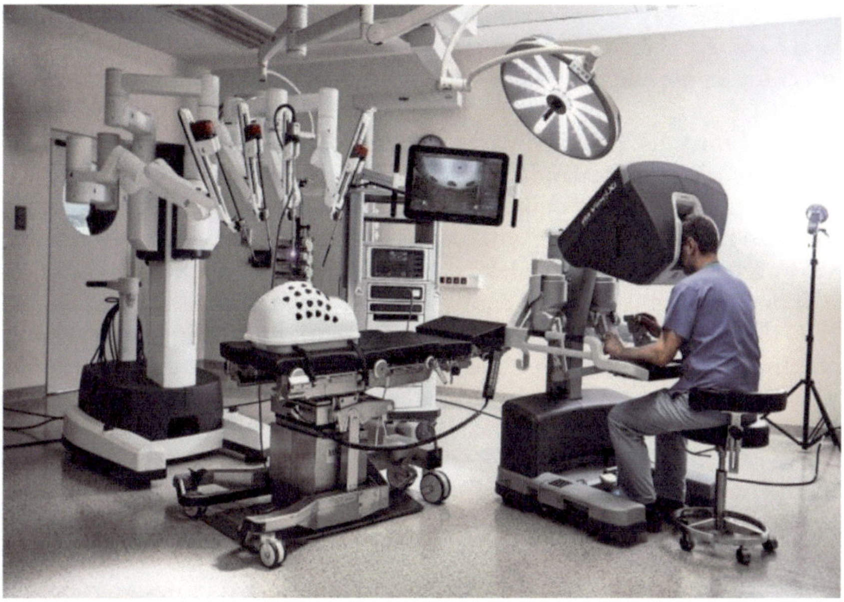

Figure 1.
Da Vinci surgical robotic system.

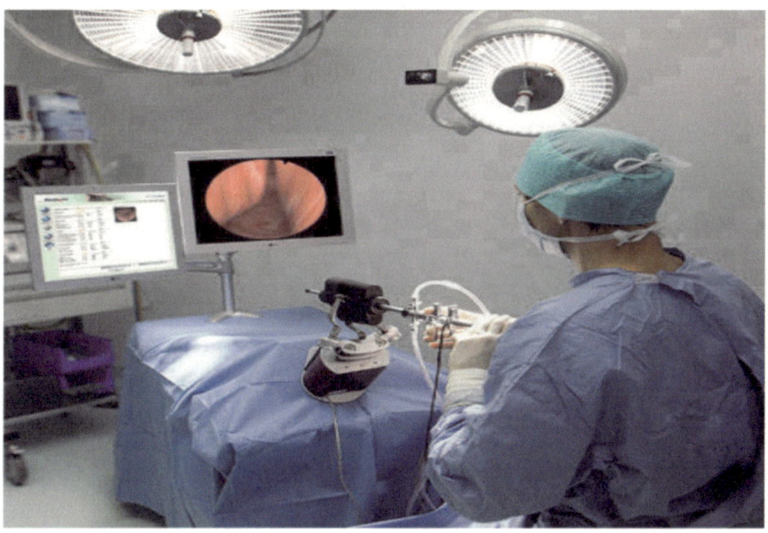

Figure 2.
URP Mentor (simulator).

feels. Such a force, generated in opposition to his movements, must be ideally the same as that the robot applies over the patient at *the slave side*. It has widely used virtual reality systems in minimally invasive surgical simulation, where the operator should feel the same forces that it would feel in a real procedure, [5].

Goertz presented the first *master-slave teleoperation system* [6]. It was used to handle toxic waste using two coupled manipulators. Subsequently, using electrical signals, he included a rudimentary force generation device. Since then, we have used this system in areas such as micro and nano-manipulation [7] underwater exploration [8], and tele-surgery [9]. Haptic feedback takes special relevance in this last area since it is crucial for the surgeon to receive an accurate force response.

The most effective way to do so is to have force sensors at both *master* and *slave* sides, and a control algorithm that guarantees an accurate tracking between the contact forces present in the remote environment and those sent to the operator [10].

The idea of force feedback on virtual reality systems began with the fundamental work of Sutherland [11]. He established that the interaction between the human operator and the virtual environment should not only be visual but also tactile. It was not until the 1990s that he adapted the Goertz device to provide force feedback during virtual molecular coupling [12]. Since then, the use of manipulators in virtual reality applications has spread to CAD/CAM assembly [13], aerospace maintenance [14], and especially in medical training through simulators [15] where, unlike systems of master-slave teleoperation, neither the environments nor the contact forces exist. It must transmit the actual forces to the operator with precision. The quality of this transmission depends on the characteristics of the haptic interface and the corresponding force control algorithm [3].

Articulated robots play a major role in medical training simulation systems with force feedback since this kind of electromechanical device can measure spatial position and generate torques. There has been a large effort to design robot haptic interfaces such as the widely used Phantom serial robot [16], the Delta Haptic parallel robot [17], and the combination of passive elements such as brakes and springs with motors [18]. Such robots are examples of impedance types devices, i.e. they read position and control force in response. There are another robots that read forces and control motion, called *admittance type devices*. The difference between using one or another type relies on the characteristics of the virtual environment (e.g., stiffness, inertia, damping, friction).

Along with haptic interfaces, there has been an intense development of graphical simulation tools capable of reproducing a wide range of virtual environments. The principal aim is for the operator to perceive, as realistic as possible, objects with a high quality of detail. The applications developed include microscopic exploration [19], aviation [20], and clinical neuropsychology [21], among many others. With medical training simulation, a correct synergy between visual and tactile feedback is essential to heighten the skills of medicine students. However, to increase immersion and consequently the realism of the virtual reality displays, it is necessary to model environments that combine haptic and graphics to the same complexity [22]. This is not always possible since it limited the computational processing and it cannot execute the applications in real-time.

Salisbury et al. [23] presented the basic architecture for a virtual reality application with visual and haptic feedback. They established that the force rendering algorithms must be geometry-dependent. This is a disadvantage in medical training simulation since the virtual objects to be reproduced include bones and organs with irregularities or indentations. There are cases in which the interaction occurs not only on the surfaces of the object, but we must also calculate the penetration forces, as we do in surgery simulators. The alternative is to design algorithms based on physical laws that involve the dynamic and movement of the objects when the operator interacts with them [24]. The perfect scenario would be to render forces by combining physical approaches with the most sophisticated haptic interfaces. However, as mentioned before, doing that is computationally more expensive, not to mention the high costs it would entail.

The more realistic the force transmitted to the operator, the higher the quality of the method used. The factors mentioned above cause a series of compensations between haptic and visual realism, real-time execution, and system costs. Such offsets allowed for establishing two principal methods for rendering forces from virtual environments [25]. The first one is *the penalty method*, which is widely used for its simplicity since it required a penetration measure starting at the contact point with the virtual object. The second approach is the *imposed motion method*, where it is considering the contact as a bilateral constraint and it is calculating the contact force response using Lagrange multipliers. Computer graphics and haptic applications use both methods. Ruspini et al. established the differences between using both methods [22].

It is to assume, according to *the penalty method*, virtual objects are formed geometrically or algebraically by defined primitives such as lines, planes, spheres, and cylinders [26]. Therefore, the force rendering depends on an implicit equation and a contact point given by a collision detection algorithm. In 1991, Sclaroff and Pentland proposed a generalization using the implicit function representation method to allow collision detection for common 3D shapes [27]. The method is to replace the polygon and spline with their previous versions. By using this technique, it is possible to use local gradients in the normal direction of the virtual surface [28]. In this sense, the concept of impedance takes special relevance since it addressed the force rendering problem as an energy exchange phenomenon, allowing to study of the stability of the haptic system [29].

Using the *imposed motion method* implies considering the contacts with the virtual object as bilateral constraints [25]. Therefore, haptic systems are a *master-slave scheme* where the rigid object, the virtual representation of the haptic interface, is *the master*. This entails employing Lagrange multipliers to compute the magnitude of the contact force. In 1994, Bayo and Avello designed an algorithm by considering the dynamics of a multi-body system as a constraint [30]. The advantage of this approach is that the force response can not only be in terms of a single contact point, but it can also be in terms of the dynamic characteristics of the surfaces [31]. However, the hardware requirement for haptic rendering in 3D is for the haptic interface to have at least 3 degrees of freedom.

The methods mentioned above have been the cornerstone for virtual forces generation both in graphic computing and haptic systems. One of the principal requirements in such areas is that the systems be capable of reproducing the forces that would be present during contact with rigid and soft objects. This is especially important in the development of simulators used in medical training, where the tactile sensations caused by contact between a virtual tool and bones or organs must reproduce [32]. Depending on the goals of the design, we must make a compromise between haptic and visual realism.

2. Preliminaries

In order to implement a virtual reality application, it is important to combine and match both visual and haptic feedback in real-time. Next, we introduce the fundamentals of virtual surface representation, which allow us to show how works the *force rendering method*, proposed in [33]. The mathematical nature of holonomic and non-holonomic constraints is presented by introducing some basic concepts of differential topology and the mathematical model of the haptic system. For this, a teloperation scheme is adapted considering the slave robot as virtual. It simulated its dynamic model within the virtual environment. Said robot is in contact with holonomic or non-holonomic virtual constraints, whose mathematical representation is also presented. The process for rendering the forces to be transmitted to the operator is detailed below, as well as some observations on the validity of the approach, especially regarding the differences when using both types of constraints.

2.1 Virtual surfaces representation

It is important to distinguish between the different representations of virtual surfaces, both from the point of view of computer graphics and from the point of view of haptic systems. Because the physics of light (with visual representation) differs from the physics of mechanical interactions. It is important to consider that although the graphical and haptic simulation can share the coding of certain properties, such as the shape, they must differ in other aspects, such as models, mathematical techniques, and implementation [32]. It is important to note that haptic rendering in practice avoids the many complex renderings developed by the graphics computing community over the years. However, in this section, the basics of such representations are given to introduce the fundamentals of the proposed haptic representation method, which central idea is to relate holonomic and non-holonomic constraints, which mathematically have a kinematic basis, with the rigid and soft tissues of the body. Surface dynamic complexity is avoided, and instead, from a purely haptic approach, it is classified as nonpenetrable and penetrable. By using the approach of a manipulator in constrained motion, it includes the dynamic of the virtual robot to exemplify that it is possible to model the virtual tool, in a more complex way than that of a single point probe used commonly in computer graphics [34].

2.1.1 Nonpenetrable virtual surfaces

In order to represent forces coming from rigid surfaces, it suffices to define algebraically an implicit equation in joint coordinates (task space), or at least two of its geometric characteristics such as normal and tangential vectors, or distance and angle relationships between points, lines, and planes [35]. This simplifies the graphical and haptic implementation since it is possible to define *the surface* as the zero set of a function f valued at $\mathbb{R}$ as $S_f = \{\boldsymbol{x} \in \mathbb{R}^3 \lfloor f(\boldsymbol{x}) = 0\}$ [36]. Based on this approach, lines, points, and mainly zero-width polygons, assembled by vertices, form rigid virtual environments, [31]. For example, in **Figure 3**, a human skull using a triangle mesh with Phys X® by NVIDIA graphics engine is modeled.

In this case, a `PxGeometry Class` by NVIDIA, part of a *geometry class*, from the *common base class*, is used. Each *geometry class* defines a volume or surface with a fixed position and orientation. We can implement it in many geometry types, as simple as spheres, capsules, boxes and planes, and others more complex as convex meshes, triangle meshes, and height fields [37]. Besides, the methods to build a complex object as a human skull and its properties include triangle mesh collision

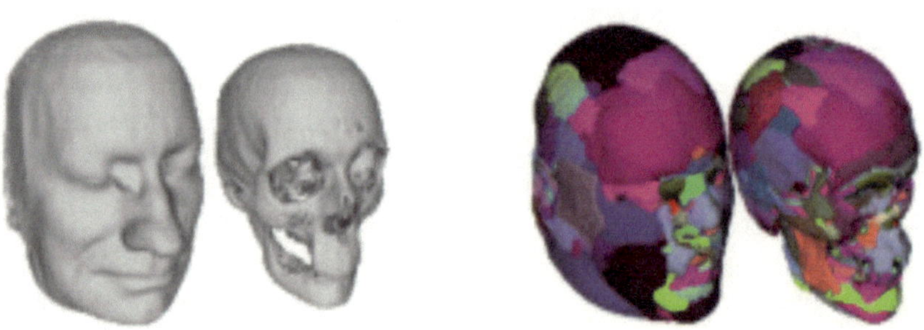

(a) (b)

Figure 3.
Convex decomposition by Phys X®. (a) Triangle mesh collision. (b) Convex decomposition.

Figure 4.
Haptic interaction with a sphere. (a) Applied external forces. (b) Collision detection. (c) Collision response. (a) Original mesh. (b) Delaunay triangulation. (c) Mass-spring like system.

and convex decomposition, **Figure 3**. The haptic interaction for virtual objects formed by such geometries and methods comes from a normal vector over the geometry, as we can see in **Figure 4**, where the haptic interaction with a sphere occurs using a collision detection algorithm of a single point.

2.1.2 Penetrable virtual surfaces

The case of deformable surfaces is more complex since, from a biomedical approach, a physically realistic simulation must consider all the nonlinearities of material deformation (e.g. stress, strain, elasticity, and viscoelasticity). One strategy is to combine a finite element discretization of the geometry together with a finite difference discretization of time and an updated Lagrangian iterative scheme [38]. Another very used representations of deformable surfaces in computer graphics are the *particle-based model*. Their location, speed, acceleration, mass, and any other parameter needed for an application describe particles and they develop according to Newtonian mechanical laws [36]. For example, in **Figure 5**, we model a deformable tissue using the graphic engine FLEX® by NVIDIA where, from a polygon-based mesh, a particle system is got through a Delaunay triangulation.

Using *particle-based models* allows to visually reproduce more complex processes such as cutting and slitting, processes very common in medical training. However, no matter how complicated the underlying model is, the force response because of deformation is a function of deflection only, that is, we define deformation as the displacement of the initial contact point between an instrument and a body. Deformable, even if it is a very large deformation, [32]. This characteristic allows us to study the contact from a kinematic perspective, it moved once the manner tool inside the soft object.

More sophisticated graphic tools such as SOFA (Simulation Open Framework Architecture) address the description of the object, typically by using three models: an internal model with independent degrees of freedom, the mass, and the constitutive laws, a collision model with contact geometry, and a visual model with detailed geometry and rendering parameters [39]. During runtime, it synchronizes the models using a generic mechanism called *mapping*, which is used to propagate forces and displacements and to enforce coherence between them. Normally, the internal model acts as the master system, imposing its displacements on the slave systems (the visual and collision model).

Let f be the function used to map the positions x_{m} of a master model to the position x_{s} of a slave model

$$\boldsymbol{x}_{\mathrm{s}} = f(\boldsymbol{x}_{\mathrm{m}}). \tag{1}$$

(a) (b) (c)

Figure 5.
Particle model by FLEX (soft tissue). (a) Original mesh. (b) Delaunay triangulation. (c) Mass-spring like system.

We map the velocities similarly to

$$\dot{\boldsymbol{x}}_{\mathrm{s}} = \boldsymbol{J}(\boldsymbol{x}_{\mathrm{m}})\dot{\boldsymbol{x}}_{\mathrm{m}} \tag{2}$$

where the Jacobian matrix $\boldsymbol{J}(\boldsymbol{x}_{\mathrm{m}})$ encodes the linear relationship between the velocities of the master and slave systems. In linear assignments, the operators f and $\boldsymbol{J}(\boldsymbol{x}_{\mathrm{m}})$ are the same, otherwise f is not linear regarding $\boldsymbol{x}_{\mathrm{m}}$, and we cannot write it as a matrix.

Given forces $\boldsymbol{\lambda}_{\mathrm{s}}$ applied to a slave model, the mapping computes the equivalent forces $\boldsymbol{\lambda}_{\mathrm{m}}$ applied to its master. Since equivalent forces must have the same energy, [39], the following relation holds

$$\dot{\boldsymbol{x}}_{\mathrm{m}}^{\mathrm{T}}\boldsymbol{\lambda}_{\mathrm{m}} = \dot{\boldsymbol{x}}_{\mathrm{s}}^{\mathrm{T}}\boldsymbol{\lambda}_{\mathrm{s}}. \tag{3}$$

The kinematic relation (2) allows to rewrite the Eq. (3) as

$$\dot{\boldsymbol{x}}_{\mathrm{m}}^{\mathrm{T}}\boldsymbol{\lambda}_{\mathrm{m}} = \boldsymbol{J}^{\mathrm{T}}(\boldsymbol{x}_{\mathrm{m}})\dot{\boldsymbol{x}}_{\mathrm{m}}^{\mathrm{T}}\boldsymbol{\lambda}_{\mathrm{s}}. \tag{4}$$

Since Eq. (4) holds for all velocities $\dot{\boldsymbol{x}}_{\mathrm{m}}$, the principle of virtual work allows us to simplify it to get

$$\boldsymbol{\lambda}_{\mathrm{m}} = \boldsymbol{J}^{\mathrm{T}}(\boldsymbol{x}_{\mathrm{m}})\boldsymbol{\lambda}_{\mathrm{s}}. \tag{5}$$

The kinematic mappings (1), (2), and (5) allow to compute displacements and to apply forces. They are also used to connect generalized coordinates, such as joint angles, to task space geometries.

2.2 Geometry of a constrained sub-manifold

From a robot control approach, we can consider a tool in contact with an object as a robot in constrained motion. The constraints of this system will be well defined if we associate them with physically realizable forces. This occurs, for example, with an industrial robot in contact with a proper surface (a real one) like a car bonnet in a painting or welding task. But with virtual environments, where surfaces do not exist, there are no physical constraint forces associated to them. Thus, the constraints are not well defined, and they are called *virtual constraints*, [40]. It mathematically addressed the nonpenetrable and penetrable virtual surfaces as virtual constraints. In this sense, it is important to introduce the geometric properties of such constraints in order to define them as either holonomic or non-holonomic.

Let Q be the n-dimensional smooth manifold configuration space of an unconstrained manipulator and $\boldsymbol{q} \in \mathbb{R}^n$ its local generalized coordinates. The tangent space to Q at $\boldsymbol{q}$, denoted $\tau_{\boldsymbol{q}}Q$ comprises all generalized velocity vectors $\dot{\boldsymbol{q}} \in \mathbb{R}^n$ of the system.

Definition 2.1. A geometric constraint on Q is a relation of the form

$$\boldsymbol{h}_i(\boldsymbol{q}) = \boldsymbol{0} \qquad i = 1, \dots, k < n, \tag{6}$$

where $\boldsymbol{h}_i : Q \to \mathbb{R}$ limits the admissible motions of the system to a $(n-k)$-dimensional smooth sub-manifold of Q ■.

Constraints that involve not only the generalized coordinates but also their first derivatives in the form

$$\boldsymbol{a}_i(\boldsymbol{q}, \dot{\boldsymbol{q}}) = \boldsymbol{0} \qquad i = 1, \dots, k < n, \tag{7}$$

with $\boldsymbol{a}_i(\boldsymbol{q}, \dot{\boldsymbol{q}}) \in \boldsymbol{\tau}_{\boldsymbol{q}}Q$, are called *kinematic constraints*. These limit the allowable movements of the manipulator to a $(n-k)$-dimensional smooth sub-manifold of Q by restricting the set of generalized velocities that can be achieved in a configuration.

Definition 2.2. A *Pfaffian constraint* on Q is a set of k kinematic constraints, which are linear in velocity in the following form

$$\boldsymbol{a}_i^{\mathrm{T}}(\boldsymbol{q}, \dot{\boldsymbol{q}}) = \boldsymbol{0} \qquad i = 1, \dots, k < n, \tag{8}$$

where $\boldsymbol{a}_i(\boldsymbol{q}, \dot{\boldsymbol{q}}) : Q \to \mathrm{R}^n$ is assumed to be smooth and linearly independent. ■

A kinematic constraint can be integrable, there are k real-valued functions $\boldsymbol{h}_i(\boldsymbol{q})$ such that

$$\frac{\partial}{\partial \boldsymbol{q}}(\boldsymbol{h}_i(\boldsymbol{q})) = \boldsymbol{a}_i^{\mathrm{T}}(\boldsymbol{q}) \qquad i = 1, \dots, k < n. \tag{9}$$

Here, the kinematic constraints are, in fact, geometric constraints. Pfaffian constraints set of $\boldsymbol{a}_i(\boldsymbol{q})$ is known as *holonomic* if it is integrable, the system has a geometric limitation. For example, by considering a set of holonomic constraints characterized by

$$\boldsymbol{\varphi}_i(\boldsymbol{q}) = \boldsymbol{0} \qquad i = 1, \dots, k < n. \tag{10}$$

From Eq. (9), we get

$$\frac{\partial}{\partial \boldsymbol{q}}(\boldsymbol{\varphi}_i(\boldsymbol{q})) = \boldsymbol{J}_{\varphi}^{\mathrm{T}}(\boldsymbol{q}) \tag{11}$$

where $\boldsymbol{J}_{\varphi}(\boldsymbol{q}) \in \mathbb{R}^{k \times n}$ is the Jacobian of the holonomic constraint. Therefore, *holonomic constraints* are characterized by equivalent equations in terms of position variables, we can get the position equations by integrating them, if velocity equations initially described the constraints, [41].

Property 2.1. Given n generalized coordinates $\boldsymbol{q}$ in a sub-manifold Q and k holonomic constraints, the space tangent to Q in a configuration can be described by adequately defining $(n-k)$ new generalized coordinates of the restricted sub-manifold that characterize the real degrees of freedom of the system [42]. ■

The set of Pfaffian constraints $\boldsymbol{a}_i(\boldsymbol{q})$ is called *non-holonomic* if it is nonintegrable, the system has a kinematic limitation. Assuming again that the vectors $\boldsymbol{a}_i : Q \to \mathbb{R}^n$ are smooth and linearly independent, the non-holonomic constraints can be expressed as

$$\boldsymbol{A}(\boldsymbol{q})\dot{\boldsymbol{q}} = \boldsymbol{0} \tag{12}$$

where $A(q) \in \mathbb{R}^{k \times n}$ is the Pfaffian matrix of non-holonomic constraints and which image space produces forces to ensure that the system does not move in those directions. The presence of these constraints limits the system mobility completely differently if compared to holonomic ones, even if its generalized velocities at each point are constrained to a $(n-k)$ dimensional sub-manifold space, it is still possible to reach any configuration in Q.

Property 2.2. Given n generalized coordinates q in a sub-manifold Q and k non-holonomic constraints, the space tangent to Q in a configuration has $(n-k)$ degrees of freedom but the number of generalized coordinates cannot be reduced, [43]. ■

Remark 2.1. We assume that *velocity-level* gives non-holonomic constraints Eq. (12), and *position-level* Eq. (10) describes holonomic constraints. In practical problems, it may describe both types of constraints as *velocity-level* equations. ■

2.2.1 Integrability of the constraints

A vector field $g : \mathbb{R}^n \to \tau_q \mathbb{R}^n$ is a smooth mapping assigning to each point $q \in \mathbb{R}^n$ a tangent vector $g(q) \in \tau_q \mathbb{R}^n$. In local coordinates, we can represent q as a column vector whose elements depend on q as

$$g(q) = \begin{bmatrix} g_1(q) \\ \vdots \\ g_n(q) \end{bmatrix}, \tag{13}$$

where g is smooth if each $g_i(q)$ is smooth.

Given g_1 and g_2, we define the *Lie bracket* of these vectors fields as

$$[g_1, g_2] = \frac{\partial}{\partial q}(g_2)g_1 - \frac{\partial}{\partial q}(g_1)g_2 \tag{14}$$

where $[g_1, g_2]$ is a new vector field.

A distribution assigns a subspace of the tangent space to each point in $\mathbb{R}^n$ smoothly. A special case is a distribution defined by a set of smooth vector fields, $g_1, \ldots, g_m$. Here, we can define distribution as

$$\Delta = \text{span}\{g_1, \ldots, g_m\}, \tag{15}$$

where the span over the set of smooth real-valued functions on $\mathbb{R}^n$ is taken. Evaluated at any point $q \in \mathbb{R}^n$, the distribution defines a linear subspace of the tangent space

$$\Delta_q = \text{span}\{g_1(q), \ldots, g_m(q)\} \subset \tau_q \mathbb{R}^n. \tag{16}$$

A distribution is *involutive* if it is closed under the Lie bracket, i.e.,

$$[g_i, g_j] \in \Delta, \qquad \forall g_i, g_j \in \Delta. \tag{17}$$

We said distribution Δ of dimension k to be integrable if, for every point $q \in \mathbb{R}^n$, there are a set of smooth functions $h_i : \mathbb{R}^n \to \mathbb{R}$ for $i = 1, \ldots, n-k$ such that the row vectors $\frac{\partial}{\partial q}(h_i)$ are linearly independent at $\mathbf{q}$ and for every $\mathbf{g} \in \Delta$.

$$\frac{\partial}{\partial q}(h_i)g(q) = \mathbf{0} \quad i = 1, \ldots, n-k. \tag{18}$$

The hypersurfaces defined by the level sets $\{\boldsymbol{h}_1(\boldsymbol{q}) = c_1, \ldots, \boldsymbol{h}_{n-k}(\boldsymbol{q}) = c_{n-k}\}$ are called *integral manifolds* for the distribution Δ. Eq. (18) shows that Δ coincides with the tangent space to its integral manifold at $\boldsymbol{q}$. We relate integral manifolds to *involutive* distributions by the following so-called Frobenious theorem, [43].

Theorem 2.1. A distribution is integrable if and only if it is involutive. ■

This theorem gives a necessary and sufficient condition for the complete integrability of a distribution. Thus, if Δ is a k-dimensional involutive distribution, then locally there are $n-k$ functions $\boldsymbol{h}_i : \mathbb{R}^n \to \mathbb{R}$ such the level surfaces $\boldsymbol{h} = (\boldsymbol{h}_1, \ldots, \boldsymbol{h}_{n-k})$ give that integral manifolds of Δ. The result mentioned above gives conditions for the integrability of a set of kinematic constraints in the following proposition.

Proposition 2.1. [44] The set of k Pfaffian constraints, described in (8), is holonomic if and only if Δ its an involutive distribution. ■

So it is possible to establish when a Pfaffian constraint is non-holonomic by checking if its distribution is not involutive.

2.3 Haptic system overview

A common practice in the computer graphics community is to associate the position and orientation of virtual tools directly with that of the haptic interface. However, this assumption is because some real tools have negligible dynamics, such as a scalpel. From a teleoperation approach, we assumed that even the simplest tool has some dynamic properties to consider in the virtual environment. This section presents a description of this proposal, both mathematical and intuitive.

In order to describe the operation of the haptic system, two independent sets of task space coordinates are considered as shown in **Figure 6**.

The operator manipulates the haptic interface, i.e., the master robot in the real environment and we denote whose Cartesian coordinates as $\boldsymbol{x}_{\mathrm{m}} \in SE(3)$, where

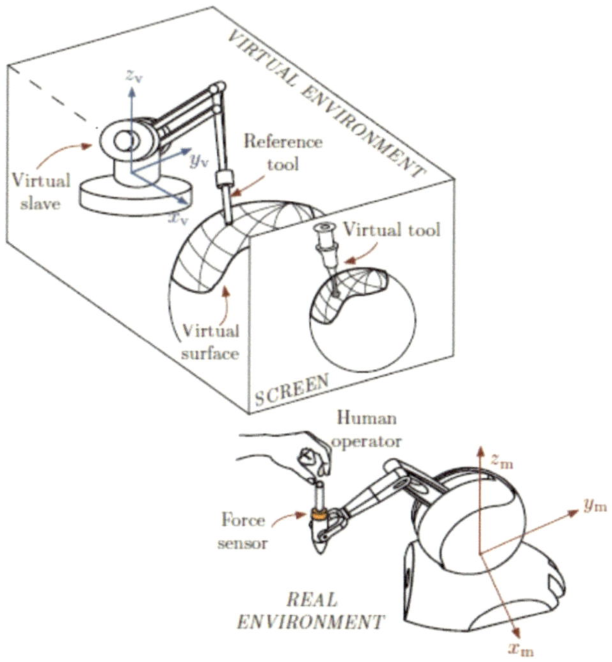

Figure 6.
Haptic system.

$\boldsymbol{p}_{\mathrm{m}} \in \mathrm{R}^3$ is the end-effector position and $\boldsymbol{R}_{\mathrm{m}} \in SO(3)$ its orientation. The virtual tool must respond to the movements of such interface in the virtual environment with Cartesian coordinates $\boldsymbol{x}_{\mathrm{v}} \in SE(3)$, where $\boldsymbol{p}_{\mathrm{v}} \in \mathbb{R}^3$ is the virtual tool position and $\boldsymbol{R}_{\mathrm{v}} \in SO(3)$ its orientation. In a teleoperation context, the position of the master robot acts as a reference for the virtual tool and it is visually projected on the screen through the virtual avatar of the system. The operator moves the virtual tool freely until the collision detection algorithm shows that contact with the virtual surface is taking place. Until now, the master robot exerts a force that is measured by a force sensor that serves as a reference for the virtual robot and must apply to that surface. By closing the feedback loop, the control algorithm produces a tactile sensation for the operator. Ideally, both visual and haptic feedback must coincide, allowing the operator to have a visual reference to the virtual tool and the feeling of the dynamic changes of its contact with the virtual surface.

In a similar approach, Faurlin et al. clarify in [44] that the virtual environment can be represented by a set of generalized coordinates $\boldsymbol{q}_{\mathrm{v}} \in \mathbb{R}^3$, which are related to the task-space coordinates of the master robot by a nonlinear kinematic equation

$$\boldsymbol{x}_{\mathrm{m}} = f(\boldsymbol{q}_{\mathrm{v}}) \tag{19}$$

which is a mapping between the real and virtual environment, similar to that of Eq. (1) but where the former acts as a master model.

The set of coordinates $\boldsymbol{q}_{\mathrm{v}}$ allows the dynamic model of the virtual tool to be described in terms of Euler-Lagrange equations of motion. A set of holonomic or non-holonomic constraints that represent the virtual surface can be embedded into the kinematic mapping (19), relating independent master robot task-space coordinates and dependent virtual robot task-space coordinates. The virtual tool moves according to the physic simulation propagated in the virtual environment coordinates and always satisfies such constraints.

2.3.1 Dynamic model and properties

Consider a *real master (m)* and a *virtual slave (v)* robot system composed of two manipulators, each of them with n degrees of freedom but not necessarily with the same kinematic configuration. Each robot spans a k-dimensional task space and, based on master/virtual devices, can be scaled to meet the desired virtual application. The master dynamics is given by

$$\boldsymbol{H}_{\mathrm{m}}(\boldsymbol{q}_{\mathrm{m}})\ddot{\boldsymbol{q}}_{\mathrm{m}} + \boldsymbol{C}_{\mathrm{m}}(\boldsymbol{q}_{\mathrm{m}}, \dot{\boldsymbol{q}}_{\mathrm{m}})\dot{\boldsymbol{q}}_{\mathrm{m}} + \boldsymbol{D}_{\mathrm{m}}\dot{\boldsymbol{q}}_{\mathrm{m}} + \boldsymbol{g}_{\mathrm{m}}(\boldsymbol{q}_{\mathrm{m}}) = \boldsymbol{\tau}_{\mathrm{m}} - \boldsymbol{\tau}_{\mathrm{h}} \tag{20}$$

while the virtual slave dynamics is modeled by

$$\boldsymbol{H}_{\mathrm{v}}(\boldsymbol{q}_{\mathrm{v}})\ddot{\boldsymbol{q}}_{\mathrm{v}} + \boldsymbol{C}_{\mathrm{v}}(\boldsymbol{q}_{\mathrm{v}}, \dot{\boldsymbol{q}}_{\mathrm{v}})\dot{\boldsymbol{q}}_{\mathrm{v}} + \boldsymbol{D}_{\mathrm{v}}\dot{\boldsymbol{q}}_{\mathrm{v}} + \boldsymbol{g}_{\mathrm{v}}(\boldsymbol{q}_{\mathrm{v}}) = \boldsymbol{\tau}_{\mathrm{v}} + \boldsymbol{\tau}_{\mathrm{s}} \tag{21}$$

where the subscripts m and v denote the real master and the virtual slave manipulators, respectively. For $i = m, v$, $\boldsymbol{q}_i \in \mathbb{R}^n$ is the vector of generalized coordinates, $\boldsymbol{H}_i(\boldsymbol{q}_i) \in \mathbb{R}^{n\times n}$ is the inertia matrix, $\boldsymbol{C}_i(\boldsymbol{q}_i, \dot{\boldsymbol{q}}_i)\dot{\boldsymbol{q}}_i \in \mathbb{R}^{n\times n}$ is the vector of Coriolis and centripetal forces, $\boldsymbol{D}_i\dot{\boldsymbol{q}}_i \in \mathbb{R}^{n\times n}$ is a diagonal matrix of viscous friction coefficients, $\boldsymbol{g}_i(\boldsymbol{q}_i) \in \mathbb{R}^n$ is the vector of gravitational torques, $\boldsymbol{\tau}_i \in \mathbb{R}^n$ is the vector of generalized inputs, $\boldsymbol{\tau}_{\mathrm{h}} \in \mathbb{R}^n$ is the real torque applied by the human operator on the master side and $\boldsymbol{\tau}_{\mathrm{s}} \in \mathbb{R}^n$ is the virtual torque generated because of the contact with the virtual constraint [45].

Property 2.3. With a proper definition of the robot parameters, it is possible to express the robot dynamics as

$$H_i(q_i)\ddot{q}_i + C_i(q_i, \dot{q}_i)\dot{q}_i + D_i\dot{q}_i + g_i(q_i) = Y_i(q_i, \dot{q}_i, \ddot{q}_i)\theta_i \tag{22}$$

where $Y_i(q_i, \dot{q}_i, \ddot{q}_i) \in \mathbb{R}^{n_i \times l}$ is the regressor and $\theta_i \in \mathbb{R}^l$ is a constant vector of parameters. ■

Assumption 2.1. The master and the virtual slave robots share the same geometric structure, but they do not necessarily have the same parameters of the dynamic model, that is, the matrices and vectors of the models described in (20) and (21) do not have to be the same. ■

External torques are acting in both robots, either the real torque τ_h applied by the human on the master side or the virtual torque τ_s generated because of the contact between the virtual robot and the virtual surface. We can define the torque applied by the human operator as

$$\tau_h = J_m^T(q_m)F_h \tag{23}$$

where $F_h \in \mathbb{R}^3$ is the force applied by the operator in the task-space coordinates and $J_m^T(q_m) \in \mathbb{R}^{3\times n}$ the geometric Jacobian of the master manipulator. In the same way, the torque applied on the virtual surface can be expressed as

$$\tau_s = J_v^T(q_v)F_s \tag{24}$$

where $F_v \in \mathbb{R}^3$ is the force applied on such surface in task-space coordinates.

2.3.2 Virtual holonomic constraints

Whit of holonomic constraints we assume that, in virtual task space coordinates, the virtual robot is subject to k virtual holonomic constraints characterized by

$$\varphi_v(x_v) = 0 \tag{25}$$

where a suitable normalization is done for the gradient of this constraint,

$$J_{\varphi x_v}(x_v) = \nabla\varphi_v(x_v) \in \mathbb{R}^{k\times n}, \tag{26}$$

to be unitary.

The representation of constraint (25) in generalized virtual coordinates is

$$\varphi_v(q_v) = 0 \tag{27}$$

where $q_v \in \mathbb{R}^n$ is the vector of the virtual robot end-effector joint coordinates. The gradient of the constraint (27) is

$$J_{\varphi v}(q_v) = \nabla\varphi_v(q_v) \in \mathbb{R}^{k\times n}. \tag{28}$$

These two gradients are related by

$$J_{\varphi x_v}(q_v) = J_{\varphi x_v}(x_v)J_v(q_v) \tag{29}$$

where $J_v(q_v) \in \mathbb{R}^k$ is the geometric Jacobian of the virtual manipulator. Hence, the torque because of the contact with the virtual surface in (1) can be defined as

$$\boldsymbol{\tau}_{\mathrm{s}} = \boldsymbol{J}_{\varphi\mathrm{v}}(\boldsymbol{q}_{\mathrm{v}})\boldsymbol{\lambda}_{\mathrm{v}} \tag{30}$$

where $\boldsymbol{\lambda}_{\mathrm{v}} \in \mathbb{R}^k$ is a vector of Lagrange multipliers that represents the virtual force applied over the surface. Then, it is possible to rewrite the whole Eq. (21) as

$$\boldsymbol{H}_{\mathrm{v}}(\boldsymbol{q}_{\mathrm{v}})\ddot{\boldsymbol{q}}_{\mathrm{v}} + \boldsymbol{C}_{\mathrm{v}}(\boldsymbol{q}_{\mathrm{v}}, \dot{\boldsymbol{q}}_{\mathrm{v}})\dot{\boldsymbol{q}}_{\mathrm{v}} + \boldsymbol{D}_{\mathrm{v}}\dot{\boldsymbol{q}}_{\mathrm{v}} + \boldsymbol{g}_{\mathrm{v}}(\boldsymbol{q}_{\mathrm{v}}) = \boldsymbol{\tau}_{\mathrm{v}} + \boldsymbol{J}_{\varphi\mathrm{v}}^{\mathrm{T}}(\boldsymbol{q}_{\mathrm{v}})\boldsymbol{\lambda}_{\mathrm{v}}. \tag{31}$$

According to **Property 2.1**, the virtual holonomic constraints (2.26) reduce the number of degrees of freedom of the virtual robot and the dimension of its configuration space to a $n - k$–dimensional sub-manifold, [43].

2.3.3 Virtual non-holonomic constraints

With non-holonomic constraints, something well known is that we cannot express them as a function of only the generalized coordinates as in (25) or (27). Instead, they are commonly expressed as Pfaffian constraints. In the present case, these constraints are written more intuitively in terms of the virtual end-effector velocities $\boldsymbol{v}_{\mathrm{v}} = \left[\dot{\boldsymbol{p}}_{\mathrm{v}} \; \boldsymbol{\omega}_{\mathrm{v}}\right]^{\mathrm{T}}$ as

$$\boldsymbol{A}_{\mathrm{v}}(\boldsymbol{x}_{\mathrm{v}})\boldsymbol{v}_{\mathrm{v}} = \mathbf{0}, \tag{32}$$

where $\dot{\boldsymbol{p}}_{\mathrm{v}} \in \mathbb{R}^3$ and $\boldsymbol{\omega}_{\mathrm{v}} \in \mathbb{R}^3$ are the linear and angular velocities of the virtual end-effector, respectively, and $\boldsymbol{A}_{\mathrm{v}}(\boldsymbol{x}_{\mathrm{v}}) \in \mathbb{R}^{k\times n}$ is the corresponding Pfaffian constraint matrix. If the dynamic equations are defined in the virtual joint-space coordinates $\boldsymbol{q}_{\mathrm{v}}$, these constraints are projected via Faurling et al. [44].

$$\boldsymbol{A}_{\mathrm{v}}(\boldsymbol{q}_{\mathrm{v}}) = \boldsymbol{A}_{\mathrm{v}}(\boldsymbol{x}_{\mathrm{v}})\boldsymbol{J}_{\mathrm{v}}(\boldsymbol{q}_{\mathrm{v}}) \tag{33}$$

Assuming that the virtual robot is subject to k velocity-level equations of non-holonomic constraints characterized by

$$\boldsymbol{A}_{\mathrm{v}}(\boldsymbol{q}_{\mathrm{v}})\dot{\boldsymbol{q}}_{\mathrm{v}} = \mathbf{0} \tag{34}$$

the torque because of the contact with the virtual environment in (21) can be expressed as

$$\boldsymbol{\tau}_{\mathrm{s}} = \boldsymbol{A}_{\mathrm{v}}^{\mathrm{T}}(\boldsymbol{q}_{\mathrm{v}})\boldsymbol{\lambda}_{\mathrm{v}} \tag{35}$$

where $\lambda_{\mathrm{v}} \in \mathbb{R}^k$ is the vector of Lagrange multipliers which determines the magnitude of the constraint forces over the virtual surface. Then, it is possible to rewrite Eq. (21) as

$$\boldsymbol{H}_{\mathrm{v}}(\boldsymbol{q}_{\mathrm{v}})\ddot{\boldsymbol{q}}_{\mathrm{v}} + \boldsymbol{C}_{\mathrm{v}}(\boldsymbol{q}_{\mathrm{v}}, \dot{\boldsymbol{q}}_{\mathrm{v}})\dot{\boldsymbol{q}}_{\mathrm{v}} + \boldsymbol{D}_{\mathrm{v}}\dot{\boldsymbol{q}}_{\mathrm{v}} + \boldsymbol{g}_{\mathrm{v}}(\boldsymbol{q}_{\mathrm{v}}) = \boldsymbol{\tau}_{\mathrm{v}} + \boldsymbol{A}_{\mathrm{v}}^{\mathrm{T}}(\boldsymbol{q}_{\mathrm{v}})\boldsymbol{\lambda}_{\mathrm{v}}. \tag{36}$$

The non-holonomic constraints reduce the number of virtual robot available degrees of freedom to an $(n - k)$-dimensional sub-manifold, but they do not reduce the dimension of its configuration space [43, 46].

3. System implementation

In this section, the theoretical and practical aspects of implementing a virtual reality system with virtual restrictions are presented. The principal aspect concerned is the design of a controller capable to perform accurate haptic feedback that makes to feel the operator to be in contact with either a penetrable or nonpenetrable virtual surface. The method for visually reproducing the *virtual tool* in contact with *the virtual objects* is presented. It is important to note that this method avoids the complexity of the virtual environments currently implemented in the simulators used in medical training. However, the basic aspects addressed to show that the virtual constraints approach can be used practically, and eventually adapted to sophisticated graphic computing tools.

3.1 Virtual environment design

As mentioned in Section 2, the important aspect to get realistic haptic feedback from a surface embedded into a virtual environment comprises defining its geometry. In **Figure 7**, we show an idealized representation of a virtual point probe in contact with either an nonpenetrable or penetrable virtual surface.

In the first case, we assume that the contact arises at a single point over the surface from where the virtual probe cannot move forward, i.e., its velocity is equal to zero. Therefore, a normal force vector, which magnitude increases depending on the force applied by the operator, avoids motion. In robot control, if we are supposed to connect the probe to the end effector of a manipulator, according to **Property 2.1**, the number of degrees of freedom of the system in contact with the surface is reduced. Intuitively, that means that the virtual probe cannot move forward from where the contact arises, which can be at any point on the surface [47]. Actually, if the virtual object is built up by using a polygonal method, there will be a set of surfaces $(\boldsymbol{\varphi}_0, \boldsymbol{\varphi}_1, \dots, \boldsymbol{\varphi}_n)$ joined by vertices as shown in **Figure 7**.

The best way to find the place of the contact point (which belongs to a set of points defining each surface) is by establishing an implicit equation $\boldsymbol{\varphi}_{\mathrm{v}}(\boldsymbol{x}_{\mathrm{v}})$ containing such a point. The set of points defining the virtual surface are

$$\boldsymbol{\varphi}_{\mathrm{v}}(\boldsymbol{x}_{\mathrm{v}}) = \mathbf{0}, \tag{37}$$

which coincides with the holonomic constraint of Eq. (25) in Cartesian coordinates or Eq. (27) in generalized coordinates. From those expressions, it can establish a collision detection algorithm by defining the following conditions:

- If $\boldsymbol{\varphi}_{\mathrm{v}}(\boldsymbol{x}_{\mathrm{v}}) > \mathbf{0}$, the virtual probe is in free motion, i.e., it is not in contact with the virtual surface.

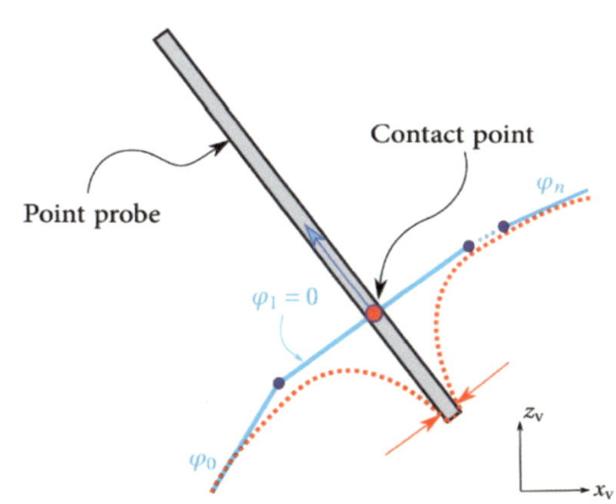

Figure 7.
(⋯) Penetrable and (__) nonpenetrable surfaces.

- If $\boldsymbol{\varphi}_{\mathrm{V}}(\boldsymbol{x}_{\mathrm{V}}) = \mathbf{0}$ the virtual probe is in contact with the virtual surface and it stays over the surface only, staying in constrained motion.

- If $\boldsymbol{\varphi}_{\mathrm{V}}(\boldsymbol{x}_{\mathrm{V}}) < \mathbf{0}$ virtual probe is in restricted movement, but we violate restriction i.e., it is inside of the virtual surface.

For the virtual reality system, it is important to remember that the vector $\boldsymbol{x}_{\mathrm{V}}$ represents the virtual robot's end-effector position. The virtual probe would share such a position. By extending the approach proposed, the virtual robot acts in fact as the virtual tool and it projected its position to *the operator* by *the avatar* of the system. In the second case, for an nonpenetrable constraint, even when the contact starts at a single point, the properties of the surface allow the virtual probe to stay in motion, as we can see in **Figure 7**. Intuitively, the process is more complex since, at a certain moment, the virtual probe must stop. In a medical context, that means the deformable tissue has a limited resistance that depends on its elastic properties. Since the aim is to get a reaction force that depends on the motion of the virtual probe once inside the deformable object, it is necessary to describe such motion properly. Unlike the non-deformable surfaces, the force does not arise normally to the surface at a single point, but lateral forces occur when the operator tries to move the virtual probe in such directions. Ideally, this would be true for any method to represent soft tissues, including finite element meshes and particle-based models. **Figure 8** shows the virtual tool motion inside a virtual object.

For easy visualization, it showed the motion in 2D but during the simulation, we must reproduce it in 3D with the aim of increase the realism of the application by improving the operator's dexterity. The contact begins in **stage A** where the virtual tool penetrates the object by following a straight trajectory, represented by a blue arrow, to reach the position in **stage B**. It can follow other trajectories, represented by dashed red lines, to reach the position of **stage D** or **stage C**. However, because of the *surrounding tissue*, *the tool* cannot move laterally because of the reaction forces (represented by the red arrows) preventing it along the trajectories. In contrast to what happens in the holonomic case, the virtual tool may stay in motion, i.e., its velocity differs from zero until the operator stops voluntarily. The process described above is like the motion of a wheeled car in 2D, which is perfectly described by non-holonomic constraints. In fact, if we add a third dimension,

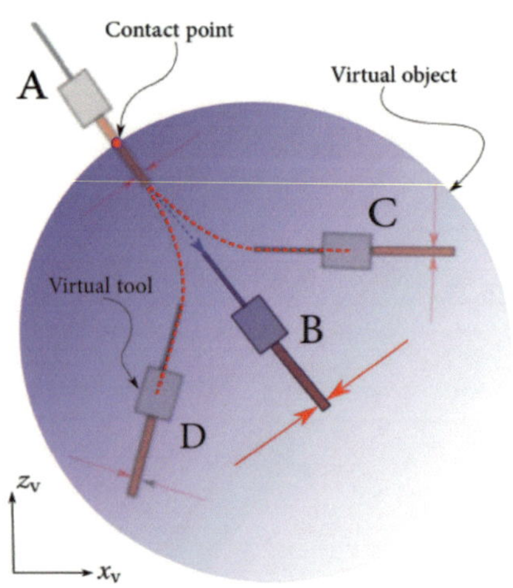

Figure 8.
Motion in 2D of the virtual tool inside a virtual object.

Property 2.2 remains forever and the *virtual tool* can reach any point on the virtual object.

The trajectories of the virtual tool shown in **Figure 8** are common in noninvasive surgical procedures. For example, in the simulator of *Transurethral Resection of the Prostate (TURP Mentor)* of **Figure 2**, the medical trainee performs a straight trajectory that, in the first place, simulates the insertion of the resectoscope into the patient's penis. Once within the virtual prostate, the resectoscope needs to be moved to remove, through an incandescent resection loop, the benign tissue that obstructs the flow of urine to the urethra. These movements follow a similar route to the one represented with red dotted lines in **Figure 8**. These movements use a pivot where the tool changes its direction.

The contribution of this approach is, in contrast to common single point haptic feedback methods, that we produce forces that prevent lateral movement of the virtual tool. However, the major disadvantage is that the environment's elastic properties are not considered. As a result, the reaction of the force that limits the movement of the operator, depending on such properties, is not reproduced and can move the *virtual tool* interchangeably within the virtual object, which does not happen in real life. For example, in human organs, elastic properties and parameters such as *Young's modulus* or *Poisson's ratio* establish limits of movement for the tool that, when exceeded, the tissue is damaged.

3.1.1 Virtual environment design

In **Figure 9**, we show a scheme in 2D of the contact between a virtual tool and a virtual surface to illustrate the use of the model given by Eq. (21).

We attach the tool to the virtual robot's last DOF, acting as its end-effector. It is important to note that the manipulator dynamic model is used to reproduce a classic bilateral teleoperation system and assuming that, since it is simulated digitally, it can be exchanged by a simpler or more complex model, including those of medical instruments such as forceps endoscopes, gripers, and retractors. This assumption leads to the proposition that, if we use the model of a surgical tool during the simulation, the realism of the contact with the surface would increase. The principal difference between defining a holonomic and a non-holonomic constraint is a need for an expression of $\varphi_v(x_v)$. Based on Section 2.1.1, from an implicit representation approach, we build rigid virtual objects from 3D basic geometric primitives as cones, pyramids, planes, cubes, and spheres, [22]. Ultimately, the base of a highly complex virtual environment composed of rigid objects is a set of basic geometric shapes that we can represent through mathematical expressions. Therefore, it is

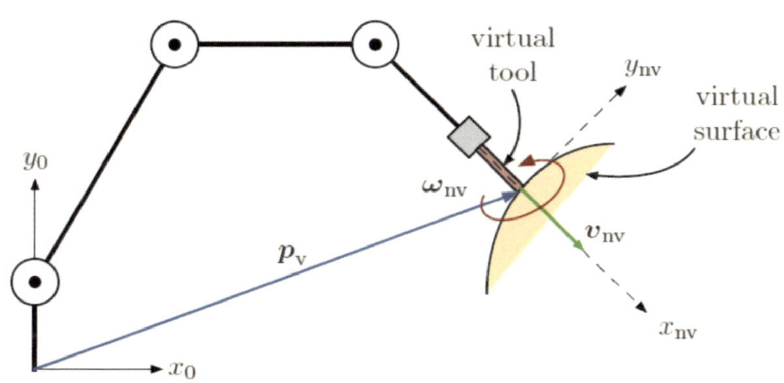

Figure 9.
Virtual robot in interaction with a penetrable surface.

enough to define a zero set of functions as in (25) that are individually expressed as $\boldsymbol{\varphi}_{\mathrm{v}}(\boldsymbol{x}_{\mathrm{v}})$ and a collision detection algorithm based on the inequalities stated above. For non-holonomic constraints and considering again the virtual robot of **Figure 9**, let ${}^{0}\boldsymbol{p}_{\mathrm{v}} \in \mathbb{R}^3$ be the Cartesian position of the virtual robot end–effector and ${}^{0}\boldsymbol{R}_{\mathrm{v}} \in SO^3$ a rotation matrix that describes its orientation. Dividing this rotation matrix into three column vectors as

$$ {}^{0}\boldsymbol{R}_{\mathrm{v}} = \left[{}^{0}\boldsymbol{x}_{n\mathrm{v}} \quad {}^{0}\boldsymbol{y}_{n\mathrm{v}} \quad {}^{0}\boldsymbol{z}_{n\mathrm{v}}, \right] \tag{38} $$

for which each column represents a vector of the end-effector coordinate frame, described in the base frame. This allows defining *Pfaffian constraints* like (29) in an intuitive form, i.e.,

$$ \boldsymbol{A}_{\mathrm{v}}\left({}^{0}\boldsymbol{x}_{n\mathrm{v}}, {}^{0}\boldsymbol{y}_{n\mathrm{v}}, {}^{0}\boldsymbol{z}_{n\mathrm{v}} \right)\boldsymbol{v}_{\mathrm{v}} = \mathbf{0}. \tag{39} $$

We claim that a set of non-holonomic constraints can be defined if the manipulator degrees of freedom are greater than those necessary to control the end-effector position, i.e., $n > 2$ for planar robots and $n > 3$ for robots in a three-dimensional workspace. The end-effector velocities of the virtual robot can be described by

$$ \boldsymbol{v}_{\mathrm{v}} = \begin{bmatrix} {}^{0}\dot{\boldsymbol{p}}_{\mathrm{v}} \\ \\ {}^{0}\boldsymbol{\omega}_{\mathrm{n}} \end{bmatrix} \tag{40} $$

where $\boldsymbol{\omega}_{\mathrm{n}}$ is the angular velocity over an axis normal to the robot plane, and ${}^{0}\dot{\boldsymbol{p}}_{\mathrm{v}}$ is the linear velocity defined as

$$ \dot{\boldsymbol{p}}_{\mathrm{v}} = \begin{bmatrix} {}^{0}\dot{p}_{\mathrm{v}x} \\ \\ {}^{0}p_{\mathrm{v}y} \end{bmatrix} \tag{41} $$

If the robot may not move in the ${}^{0}\boldsymbol{y}_{n\mathrm{v}}$ direction, the corresponding *Pfaffian constraint* is given by

$$ \left[{}^{0}\boldsymbol{y}_{n\mathrm{v}}^{\mathrm{T}} \quad 0 \right]\boldsymbol{v}_{\mathrm{v}} = \left[-\sin\left(q_{\mathrm{v}1} + q_{\mathrm{v}2} + q_{\mathrm{v}3}\right) \quad \cos\left(q_{\mathrm{v}1} + q_{\mathrm{v}2} + q_{\mathrm{v}3}\right) \quad 0 \right] \begin{bmatrix} {}^{0}\dot{p}_{\mathrm{v}x} \\ {}^{0}\dot{p}_{\mathrm{v}y} \\ {}^{0}\dot{\omega}_{n} \end{bmatrix}. \tag{42} $$

By choosing the distribution

$$ \Delta = \left[g_1, g_2 \right] \tag{43} $$

where

$$ g_1 = \begin{bmatrix} \cos\left(q_{\mathrm{v}1} + q_{\mathrm{v}2} + q_{\mathrm{v}3}\right) \\ \sin\left(q_{\mathrm{v}1} + q_{\mathrm{v}2} + q_{\mathrm{v}3}\right) \\ 0 \end{bmatrix} \tag{44} $$

$$ g_2 = \begin{bmatrix} 0 \\ 0 \\ 1 \end{bmatrix} \tag{45} $$

as a basis for the null-space of the *Pfaffian matrix*, the equivalent control system

$$\dot{q}_v = g_1 u_1 + g_2 u_2 \tag{46}$$

can be constructed, representing the directions of allowed motion, [42].

It is easy to verify that the Lie bracket is

$$[g_1 \quad g_2] = \begin{bmatrix} -\sin(q_{v1} + q_{v2} + q_{v3}) \\ \cos(q_{v1} + q_{v2} + q_{v3}) \\ 0 \end{bmatrix}, \tag{47}$$

which shows the non-involutivity of the distribution and thus establishes the non-holonomic nature of the constraints according to **Proposition 2.1**.

Notice also that if the degrees of freedom were 2, the null-space would be of dimension 1, which is necessarily involutive, and the constraints would be holonomic.

3.2 Position-force controllers design

A correct haptic rendering largely depends on the force control algorithm. In classic haptic systems, the common solution is to define indirect impedance or compliance control schemes. In contrast, in this section, we present two hybrid control algorithms for haptic interaction with virtual constrained systems. As mentioned in Section 2, the usual practice is to associate the position of the haptic interface directly with that of the virtual avatar. Therefore, a position control scheme is unnecessary, as the operator's movements are reflected in the graphical application accurately. However, in the proposed approach, the task space coordinates of the virtual environment depend on the correct tracking between the position of the haptic robot and that of the virtual one, i.e., the control algorithm generates the virtual environment itself. This is due to including the virtual robot dynamics and the fact that the operator should feel the virtual tool because of the masking effect. In order to address this, we explored a control scheme used in teleoperation to achieve both position and force tracking. Next, we show a block diagram of this scheme in **Figure 10**.

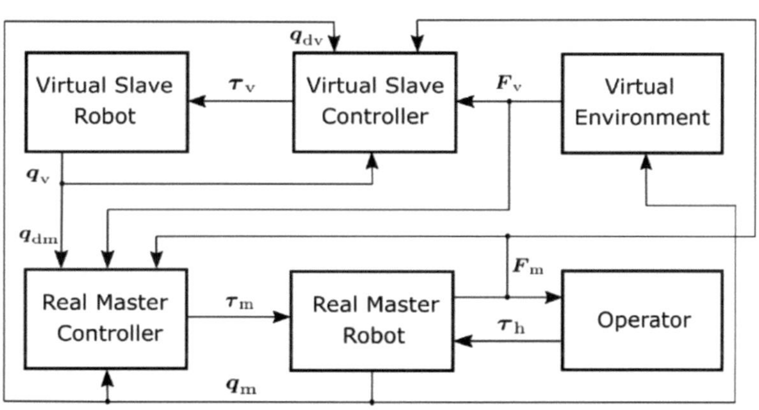

Figure 10.
Block diagram of the proposed scheme.

Considering once again $i,j = \mathrm{m}, \mathrm{v}$ where $i \neq j$. We define

$$\boldsymbol{q}_{\mathrm{d}i}(t) \triangleq \boldsymbol{q}_j(t) \tag{48}$$

as the desired position trajectories, and

$$\dot{\boldsymbol{q}}_{\mathrm{d}i}(t) \triangleq \dot{\boldsymbol{q}}_j(t) \tag{49}$$

as the desired velocity trajectories, i.e., if $i = \mathrm{m}$, then $j = \mathrm{v}$ and vice versa. We define the corresponding tracking error as

$$\Delta\boldsymbol{q}_i \triangleq \boldsymbol{q}_i - \boldsymbol{q}_{\mathrm{d}i}. \tag{50}$$

Based on [48], we proposed

$$\boldsymbol{s}_i = \dot{\boldsymbol{q}}_i - \dot{\boldsymbol{q}}_{\mathrm{d}i} + \boldsymbol{\Lambda}_{xi}\Delta\boldsymbol{q}_i. \tag{51}$$

and

$$\dot{\boldsymbol{\sigma}}_i = \boldsymbol{K}_{\beta i}\boldsymbol{s}_i + \mathrm{sign}(\boldsymbol{s}_i), \tag{52}$$

where $\boldsymbol{K}_{\beta i} \in \mathbb{R}^{n \times n}$ is a positive definite diagonal matrix and

$$\mathrm{sign}(\boldsymbol{s}_i) = \begin{bmatrix} \mathrm{sign}(\boldsymbol{s}_{i1}) \\ \vdots \\ \mathrm{sign}(\boldsymbol{s}_{in}) \end{bmatrix} \tag{53}$$

with s_{ij} element of $\boldsymbol{s}_i$ for $j = 1, \dots, n$.
Now, by considering the velocity reference as

$$\dot{\boldsymbol{q}}_{\mathrm{r}i} = \dot{\boldsymbol{q}}_{\mathrm{r}i} + \boldsymbol{\Lambda}_{xi}\Delta\boldsymbol{q}_i - \boldsymbol{K}_{\gamma i}\boldsymbol{\sigma}_i, \tag{54}$$

where $\boldsymbol{K}_{\gamma i}\boldsymbol{\sigma}_i \in \mathbb{R}^{n \times n}$ is a positive definite diagonal matrix.
We define also the auxiliary variable

$$\boldsymbol{s}_{\mathrm{a}i} = \dot{\boldsymbol{q}}_i - \dot{\boldsymbol{q}}_{\mathrm{r}i}. \tag{55}$$

By supposing that both robots are in free movement, for that case, the control laws for the master and the virtual robots are proposed as

$$\boldsymbol{\tau}_{\mathrm{m}} = -\boldsymbol{K}_{\mathrm{am}}\dot{\boldsymbol{q}}_{\mathrm{m}} - \boldsymbol{K}_{\mathrm{pm}}\boldsymbol{s}_{\mathrm{am}} \tag{56}$$

$$\boldsymbol{\tau}_{\mathrm{v}} = \boldsymbol{K}_{\mathrm{av}}\dot{\boldsymbol{q}}_{\mathrm{v}} + \boldsymbol{K}_{\mathrm{pv}}\boldsymbol{s}_{\mathrm{av}}, \tag{57}$$

respectively, where $\boldsymbol{K}_{\mathrm{am}}, \boldsymbol{K}_{\mathrm{pm}}, \boldsymbol{K}_{\mathrm{av}}$ and $\boldsymbol{K}_{\mathrm{pv}}$ are positive definite diagonal matrices.

3.2.1 Virtual holonomic constraints

Making an approximation of what happens during the tactile interaction of a point probe with a rigid surface, we considered the one-dimensional case ($\varphi_{\mathrm{v}} : \mathbb{R}^n \to \mathbb{R}$). As mentioned in Section 3.1, that is ideally the normal force generated at a single point of contact where other reactions, as friction or tangential forces, we can omit them, [32]. In order to reproduce this effect, we use the implicit surface method, which $\lambda_{\mathrm{v}} = \lambda_{\mathrm{v}}$ represents the normal force of the virtual manipulator over the virtual surface. To reflect such contact force, the Generic Penalty Method computed a Lagrange multiplier as used by [49], i.e.,

$$\lambda_{\mathrm{v}} = \alpha_{\mathrm{v}}\left(\ddot{\boldsymbol{\varphi}}_{\mathrm{v}}\left(\boldsymbol{q}_{\mathrm{v}}\right) + 2\xi\omega_{\mathrm{n}}\dot{\boldsymbol{\varphi}}_{\mathrm{v}}\left(\boldsymbol{q}_{\mathrm{v}}\right) + \omega_{n}^{2}\boldsymbol{\varphi}_{\mathrm{v}}\left(\boldsymbol{q}_{\mathrm{v}}\right)\right) \tag{58}$$

where $\xi, \omega_{\mathrm{n}} > 0$. Considering that force measurements are available at the master side in Cartesian coordinates and mapping the virtual force to this space as

$$\boldsymbol{F}_{\mathrm{v}} = \boldsymbol{J}_{\varphi x\mathrm{v}}^{\mathrm{T}}(\boldsymbol{x}_{\mathrm{v}})\lambda_{\mathrm{v}}, \tag{59}$$

we use a PID-like controller for the virtual reality system. Consider that $\boldsymbol{F}_{\mathrm{h}} \in \mathbb{R}^{3}$ is the normal force component measured with a force sensor mounted at the master robot end-effector. After (23), (24), and (58), we can use a PI controller for the virtual reality system.

By defining

$$\boldsymbol{F}_{\mathrm{d}i}(t) \triangleq \boldsymbol{F}_{i}(t) \tag{60}$$

as the desired force trajectory where if $i = \mathrm{h}$ then $j = \mathrm{v}$ and vice versa, as stated before.

The force tracking errors are

$$\Delta\boldsymbol{F}_{i} = \boldsymbol{F}_{i} - \boldsymbol{F}_{\mathrm{d}i} \tag{61}$$

and the corresponding integral, the momenta tracking error, is

$$\Delta\boldsymbol{\rho}_{i} = \int_{0}^{t}\Delta\boldsymbol{F}_{i}\mathrm{dt}. \tag{62}$$

Note that we use the standard notation for momenta $\boldsymbol{\rho}$, although also the same notation is for the position. We claim that there is no confusion because it always appears $\Delta\boldsymbol{\rho}$ in that case. Instead of using Eqs. (56) and (57) to describe the master and the virtual robot, we gave the corresponding control laws by

$$\boldsymbol{\tau}_{\mathrm{m}} = \boldsymbol{Y}_{\mathrm{m}}\left(\boldsymbol{q}_{\mathrm{m}}, \dot{\boldsymbol{q}}_{\mathrm{m}}, \ddot{\boldsymbol{q}}_{\mathrm{m}}\right)\boldsymbol{\theta}_{\mathrm{m}} - \boldsymbol{K}_{\mathrm{am}}\dot{\boldsymbol{q}}_{\mathrm{m}} - \boldsymbol{K}_{\mathrm{pm}}\boldsymbol{s}_{\mathrm{am}} + \boldsymbol{J}_{\mathrm{m}}^{\mathrm{T}}\left(\boldsymbol{q}_{\mathrm{m}}\right)\left(\boldsymbol{F}_{\mathrm{v}} - \boldsymbol{K}_{\mathrm{fm}}\Delta\boldsymbol{\rho}_{\mathrm{h}}\right) \tag{63}$$

$$\boldsymbol{\tau}_{\mathrm{v}} = \boldsymbol{K}_{\mathrm{av}}\dot{\boldsymbol{q}}_{\mathrm{v}} + \boldsymbol{K}_{\mathrm{pv}}\boldsymbol{s}_{\mathrm{av}} - \boldsymbol{J}_{\mathrm{v}}^{\mathrm{T}}\left(\boldsymbol{q}_{\mathrm{v}}\right)\left(\boldsymbol{F}_{\mathrm{h}} - \boldsymbol{K}_{\mathrm{fv}}\Delta\boldsymbol{\rho}_{\mathrm{v}}\right), \tag{64}$$

respectively, where $\boldsymbol{K}_{\mathrm{f}i} \in \mathbb{R}^{n\times n}$ are diagonal matrices. In order for the operator to feel the virtual tool in contact with the virtual environment, we can carry out a dynamic cancelation of the master manipulator dynamics, as shown in (63).

3.2.2 Virtual non-holonomic constraints

In contrast with the holonomic case, when the constraints are non-holonomic, we cannot define them as a function of a set of generalized coordinates, as stated by the Frobenius theorem. As a result, we cannot compute the Lagrange multipliers as in (58). We define these constraints in the form (31) or equivalently (34). One problem arising from these constraints is how to compute the Lagrangian multipliers to satisfy (36). These multipliers represent the forces required to maintain such constraints. Unfortunately, most of the methods used to calculate the lagrange multipliers are designed for systems with holonomic constraints [30, 49, 50] and, therefore, these methods require a position-level definition of the Pfaffian constraints as in (25) or (27). As stated in (45), the calculation presented in [42] can be used for this case. However, it is well known that this solution is unstable since its underlying mechanism is a second-order integrator with zero input. In this work, a modification of the approach used in [50] is proposed as follows. For simplicity, we define

$$\boldsymbol{H}_{\mathrm{v}} = \boldsymbol{H}_{\mathrm{v}}(\boldsymbol{q}_{\mathrm{v}}) \tag{65}$$

$$\boldsymbol{C}_{\mathrm{v}} = \boldsymbol{C}_{\mathrm{v}}(\boldsymbol{q}_{\mathrm{v}}, \dot{\boldsymbol{q}}_{\mathrm{v}}) \tag{66}$$

$$\boldsymbol{g}_{\mathrm{v}} = \boldsymbol{g}_{\mathrm{v}}(\boldsymbol{q}_{\mathrm{v}}) \tag{67}$$

$$\boldsymbol{A}_{\mathrm{v}} = \boldsymbol{A}_{\mathrm{v}}(\boldsymbol{q}_{\mathrm{v}}) \tag{68}$$

$$\boldsymbol{\psi}_{\mathrm{v}} = \boldsymbol{\psi}_{\mathrm{v}}(\boldsymbol{q}_{\mathrm{v}}, \dot{\boldsymbol{q}}_{\mathrm{v}}) = \boldsymbol{A}_{\mathrm{v}}(\boldsymbol{q}_{\mathrm{v}})\dot{\boldsymbol{q}}_{\mathrm{v}} \tag{69}$$

Then, the Lagrange multipliers can be computed as

$$\boldsymbol{\lambda}_{\mathrm{v}} = \left(\boldsymbol{A}_{\mathrm{v}}\boldsymbol{H}_{\mathrm{v}}^{-1}\boldsymbol{A}_{\mathrm{v}}^{\mathrm{T}}\right)^{-1}\left[\dot{\boldsymbol{\psi}} - \dot{\boldsymbol{A}}_{\mathrm{v}}\dot{\boldsymbol{q}}_{\mathrm{v}} - \boldsymbol{A}_{\mathrm{v}}\boldsymbol{H}_{\mathrm{v}}^{-1}(\boldsymbol{\tau}_{\mathrm{v}} - \boldsymbol{C}_{\mathrm{v}}\dot{\boldsymbol{q}}_{\mathrm{v}} - \boldsymbol{D}_{\mathrm{v}}\dot{\boldsymbol{q}}_{\mathrm{v}} - \boldsymbol{g}_{\mathrm{v}})\right], \tag{70}$$

where the constraints are forced to satisfy

$$\dot{\boldsymbol{\psi}} + 2\alpha_{\mathrm{v}}\boldsymbol{\psi}_{\mathrm{v}} + \beta_{\mathrm{v}}\int_{t_0}^{t}\boldsymbol{\psi}_{\mathrm{v}}\mathrm{d}\vartheta = \boldsymbol{0} \tag{71}$$

with $\alpha_{\mathrm{v}}, \beta_{\mathrm{v}} > 0$ chosen to ensure rapid convergence to the origin.

Note that the constraint function $\boldsymbol{\psi}$ can be defined in terms of the velocities of the end-effector,

$$\boldsymbol{\psi} = \boldsymbol{\psi}(\boldsymbol{x}_{\mathrm{v}}, \boldsymbol{v}_{\mathrm{v}}) = \boldsymbol{A}_{\mathrm{v}}(\boldsymbol{x}_{\mathrm{v}})\boldsymbol{v}_{\mathrm{v}}. \tag{72}$$

Therefore, the initial condition of the integral term on the left-hand side of (72) can be set to zero. Each element $\boldsymbol{\lambda}_{\mathrm{v}}$ is a function of $\boldsymbol{q}_{\mathrm{v}}, \dot{\boldsymbol{q}}_{\mathrm{v}}$ and $\boldsymbol{\tau}_{\mathrm{v}}$ since the constraints change with the configuration, velocity and virtual applied force.

By substituting (71) in the motion Eq. (72), a complete description of the dynamics of the system is gotten. Regarding force, sensor measurements $\boldsymbol{F}_{\mathrm{v}}$ on the master side can calculate the real Lagrange multiplier as

$$\boldsymbol{\lambda}_{\mathrm{m}} = \left(\boldsymbol{A}_{\mathrm{v}}\boldsymbol{A}_{\mathrm{v}}^{\mathrm{T}}\right)^{-1}\boldsymbol{A}_{\mathrm{v}}\boldsymbol{F}_{\mathrm{h}}. \tag{73}$$

By defining

$$\boldsymbol{\lambda}_{\mathrm{d}i}(t) \triangleq \boldsymbol{\lambda}_{\mathrm{j}}(t) \tag{74}$$

as the desired force trajectory in joint space. The corresponding integral is

$$\Delta\boldsymbol{\lambda}_{i} = \int_{0}^{t}(\boldsymbol{\lambda}_{i} - \boldsymbol{\lambda}_{\mathrm{d}j}\mathrm{d}t) \tag{75}$$

Finally, instead of (56) and (57), for the master and virtual robot the proposed position-force control for a virtual dynamic system subject to non-holonomic constraints is

$$\boldsymbol{\tau}_{\mathrm{m}} = \boldsymbol{Y}_{\mathrm{m}}(\boldsymbol{q}_{\mathrm{m}}, \dot{\boldsymbol{q}}_{\mathrm{m}}, \ddot{\boldsymbol{q}}_{\mathrm{m}})\boldsymbol{\theta}_{\mathrm{m}} - \boldsymbol{K}_{\mathrm{am}}\dot{\boldsymbol{q}}_{\mathrm{m}} - \boldsymbol{K}_{\mathrm{pm}}\boldsymbol{s}_{\mathrm{am}} + \boldsymbol{A}_{\mathrm{v}}^{\mathrm{T}}(\boldsymbol{q}_{\mathrm{v}})(\boldsymbol{\lambda}_{\mathrm{v}} - \boldsymbol{K}_{\mathrm{fm}}\Delta\boldsymbol{\lambda}_{\mathrm{m}}) \tag{76}$$

$$\boldsymbol{\tau}_{\mathrm{vm}} = \boldsymbol{K}_{\mathrm{av}}\dot{\boldsymbol{q}}_{\mathrm{v}} + \boldsymbol{K}_{\mathrm{pv}}\boldsymbol{s}_{\mathrm{av}} - \boldsymbol{A}_{\mathrm{v}}^{\mathrm{T}}(\boldsymbol{q}_{\mathrm{v}})(\boldsymbol{\lambda}_{\mathrm{m}} - \boldsymbol{K}_{\mathrm{fv}}\Delta\boldsymbol{\lambda}_{\mathrm{v}}) \tag{77}$$

Note that the novelty of the approach is not the control scheme because very well-known techniques are employed, but the novelty lies in the effective use of non-holonomic constraints to describe penetrable virtual surfaces. Therefore, a technical stability proof is not provided, but it shows a set of reliable experiments in the next section with the aim of validating the proposed approach.

3.3 Visual components of the virtual environment

A fundamental part of the developed virtual reality system is visual feedback. In dynamic systems and control research, there is no interest in including such elements but in real-world applications, as surgery simulators, it is essential. Nowadays, in those developments, we compose the virtual environments by merging several numeric techniques that, combined with the fast velocity of today's processors, give the virtual objects and surfaces a realism that before would seem impossible to reach., since the goal of this dissertation is to show how a teleoperation control scheme can be used in a virtual reality system, we design the environment by using the fundamentals of graphic computing. The tool used to design the virtual environment was the graphic standard OpenGL 2.0 which is an API, which is a software library for accessing features in graphics hardware. It contains different commands that are used to specify objects, images, and operations needed to produce interactive three-dimensional graphic applications, [51]. Among those operations, the possibility to give texture[1] and lighting to the virtual objects is possible, besides proportioning position and orientation changes to the scene's camera, i.e., the way the operator sees the images on the computer screen regarding height, deep, viewing angles as pitch, roll, and yaw, etc. As we can see in **Figure 11**, the environment of the developed application comprises a motionless floating sphere and the virtual avatar of the system.

For simplicity's sake, there are no changes in the camera's position and orientation, but we gave lighting and texture to the scene. We appreciate a notable

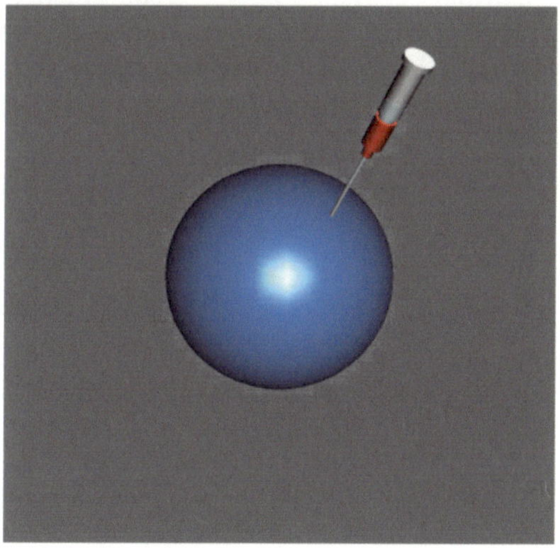

Figure 11.
Virtual environment developed in OpenGL 2.0.

[1] In graphic computing, we refer texture to the feature of give color or combinations of colors to the objects.

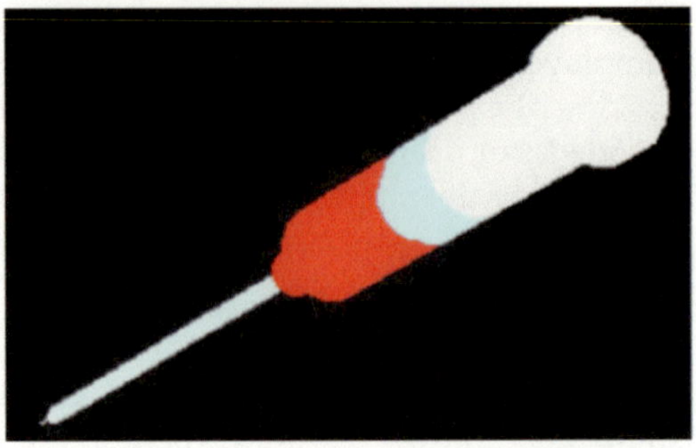

Figure 12.
Virtual avatar of the system.

difference in **Figure 12**, where the avatar has no lighting, the quality of its texture is less and the background color changes.

Here, a set of aligned cylinders becomes the avatar, which directly related position and orientation to those of the end-effector of the haptic interface. Since OpenGL has the instructions to create elements from primitives, the generation of such cylinders and the floating sphere was straightforward.

3.3.1 Rigid sphere

It is important to remember that, from a haptic rendering approach, we established the classification of nonpenetrable and penetrable objects for rigid and deformable objects, respectively. With the rigid object, the Open GL instruction `glutSolidSphere()` build automatically a solid sphere with a specific radius by defining the number of subdivisions around (sectors) and along (stacks) the z axis, as we can see in **Figure 13**.

We give the effect of rigidity because that vertex's position is not changed when contact with the virtual avatar occurs. However, giving the haptic effect of highly rigid objects to the operator was difficult since an impedance device was used. For such reason, it is important to establish the control scheme (63) and (64) that compensates, as possible, the limitations of hardware and make feel it produced contact with a rigid object.

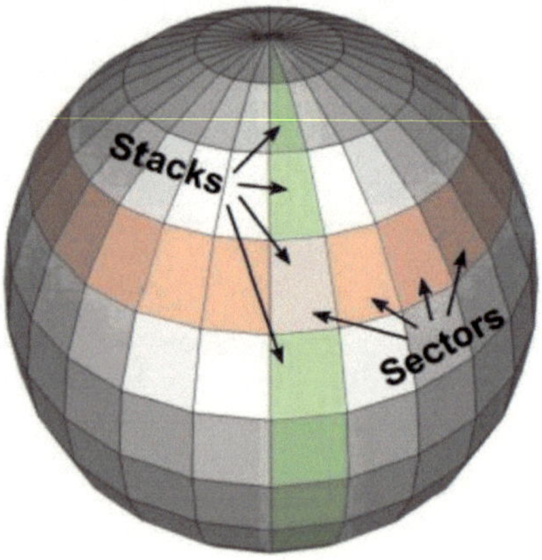

Figure 13.
Sectors and stacks of a solid sphere.

3.3.2 Deformable sphere

The real challenge comes when the *virtual object is deformable*. Saying that an object is deformable has many implications, mainly related to mechanics. Therefore, the visual effect of deformation is more complex than that produced by stiffness, because a real deformable object has an infinite number of degrees of freedom. For this reason, virtual objects need a high resolution, which gives a better rendering quality to visual and haptic feedback [52] and more realism to the application. However, this is always limited to the computational resources available. We must run both the graphics and the control algorithm in a single program, with the smallest sample time that the system allows. If the graphics part occupies more processing resources, this sampling time will increase and there will be unwanted effects, delays, and, finally, an application crash. We drew the sphere by defining four vertices a, b, c and d, which form a plane that is replicated iteratively according to several parallels p and meridians m defined by the operator. We intrinsically linked the value of the iteration to resolving the sphere. **Figure 14**, it showed the deformable sphere with different resolutions. In the sphere on the left, the values used were $p = m = 50$ while in the central sphere were $p = m = 100$. For the sphere on the right, the values were $p = m = 150$.

As mentioned before, the more the resolution of the sphere, the more the realism of the application. However, the computational processing when using the resolution of the last case did not allow a correct performance of the graphic or haptic part. For this reason, a resolution $p = m = 100$ was chosen. For simplicity's sake, the sphere was built by placing two hemispheres, one above the other, regarding a common axis. The contact with the virtual avatar will arise in a single point (x_v, y_v, z_v) computed parametrically as

$$x_v = r\cos(\alpha)\cos(\beta) \quad (78)$$

$$y_v = r\cos(\alpha)\sin(\beta) \quad (79)$$

$$z_v = r\sin(\alpha) \quad (80)$$

where r is the radius of the sphere, and α and β are the angles from whose ranges parallel and meridians are drawn. For code optimization, meridians have the range of $[0, 360]$ and parallels of $[0, 180]$. Such ranges correspond to the upper hemisphere while the lower is drawn by considering the z_v axis negative part.

Every vertex a, b, c and d must take the value of Eqs. (78)–(80) in order to visualize its initial position in the virtual environment. To improve the interaction with the virtual avatar, we include an offset r_{off} in the equations that define the vertices as

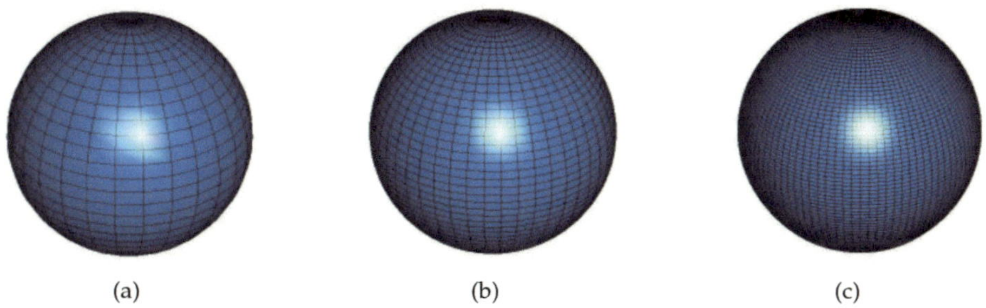

Figure 14.
Surface mesh generated using different values for p and m. (a) p = m = 50. (b) p = m = 100. (c) p = m = 150.

$$v_{nx} = (r - r_{\text{off}}) \cos(\alpha) \cos(\beta) \tag{81}$$

$$v_{ny} = (r - r_{\text{off}}) \cos(\alpha) \sin(\beta) \tag{82}$$

$$v_{nz} = (r - r_{\text{off}}) \sin(\alpha) \tag{83}$$

where $n = a, b, c, d$ and r_{off} take an arbitrary value defined experimentally. The contact with the virtual avatar occurs in some plane defined iteratively by (81)–(83) using a collision detection algorithm consisting on validating the value of each vertex of every plane and comparing the value of each component (x_v, y_v, z_v) with the position of the master robot $\boldsymbol{p}_m$. If such values belong to the range of the plane, then the avatar is in contact with the sphere. The next step is to produce the effect of motion of the contact plane. Algorithm 1 shows the pseudocode of such a process, which uses an auxiliary normal force in which direction the plane will move. This force does not belong to that calculated using the non-holonomic constraint but the gradient of Eq. (29) and we use it only for visual effect. The process described above makes only one plane to move and, if that happens, the deformation effect is not realistic. For this reason, moving the adjacent planes is necessary. It moved the more planes, the more realistic the effect of the object deformation. However, in the application developed, we changed only the position of the surrounding planes to the contact plane, since the method is purely geometric and not based on continuum mechanics. Using such an approach implies a lot of considerations that are beyond this research.

Algorithm 1 shows the part of the pseudocode where the modification of the adjacent planes takes place. This does not occur at the same time as the modification of the contact plane, since the code has not yet created these. **Figure 15** shows the visual effect that the motion, both of the contact plane and adjacent planes produce. It is important to note that the code implemented needs to be optimized and, above all, to be adapted to the force rendering algorithm through non-holonomic constraints.

4. Experimental platform

The experimental platform comprises a Geomagic Touch haptic robot with six revolute joints, having only the first three of them actuated. An ATI Nano–17 six-axis force sensor is adapted at the last link, as shown in **Figure 16**. A PC executes the control loop with a sample time of $T = 2$ milliseconds.

As mentioned in Section 3, the virtual environment comprises a sphere developed using the graphic standard OpenGL 2.0. We should note that both the control

(a) (b)

Figure 15.
Deformation effect of the sphere. (a) Contact. (b) Deformation effect.

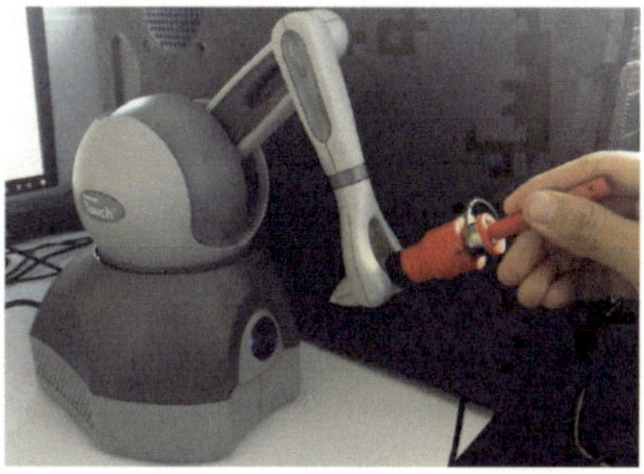

Figure 16.
Experimental platform.

algorithm and the graphic simulation run in the same application developed in Visual Studio/C++.

A practical limitation of the Geomagic Touch robot is that it actuated only the first three joints. Therefore, a projection of both the force reflection and the controller torques is necessary, i.e., the contribution of the last two joint torques is neglected. The virtual robot does not have this limitation, and therefore is considered to be fully actuated. The master robot limitation is not so restrictive, since the virtual environment considers only force but not end-effector torque feedback, avoiding the problem of sensor/actuators asymmetry haptic interfaces [53]. The contribution of the last two joints to the force reflection is much less in magnitude when compared with the contribution of the first three joints.

4.1 Task description

A detailed description of the interaction process between the virtual tool and the virtual environment is presented, simulating separately a rigid and a penetrable sphere. Since the goal of this research was to extend the use of the control scheme to medical training applications, we adopted the shape of the avatar as a *needle*, as we can see in **Figure 11**. In medicine, procedures that use this tool are very common, with *needle insertion* being the most studied and simulated procedure [54]. In this procedure, the operator takes a sterile needle and slowly brings it closer to the patient, once in contact, the operator must be very careful and, through a tactile sensation, know if the soft tissue (muscle, organ) or rigid (bone) has been affected. In both cases, the contact surface produces a reaction force in opposition to the operator's movements.

While on a rigid surface, the force does not let the needle penetrate the tissue, in a penetrable surface this is possible. The force behaviors are different, as in the first case, there is a major contribution in the normal direction, which would allow the operator to move the needle laterally over the surface. In the second case, the normal force contribution is smaller and the surrounding tissue would not allow moving the needle in the lateral directions.

In the approach presented, we assume we attach the virtual tool to the end-effector of a five degrees-of-freedom manipulator, which is not visible in the graphic simulation. It may seem counterintuitive because, evidently in real life, a needle does not have such dynamics. We use the robot model as a demonstrative example of other medical tools such as an endoscope, resectoscope, forceps. Attached to teleoperated surgical robot arm have such complex dynamics that must be modeled. The graphic simulations in those cases include pulling, gripping, clamping, and cutting, and therefore it is convenient to have a complete description of both the kinematics and the dynamics of the tool-tissue interaction, [3]. The task starts with the Geomagic Touch robot in its home position. The operator grasps the master robot stylus using the force sensor adapter tip to later gently bring it closer to the virtual surface. We imposed the desired trajectory in free motion in this way. The virtual robot moves following such a trajectory in the virtual environment, with no scaling between the virtual and the real workspaces. It perceived visually both the avatar movement and the virtual surface through a computer screen. When the collision-detection algorithm detects contact with the surface, the force-response algorithm generates a virtual force trajectory by computing the corresponding Lagrangian multipliers, either by employing (58) or (70). The operator perceives an interaction force exerted by the master robot and registered by the Nano-17 force sensor until the contact is over. Finally, the operator returns the sensor adapter to its initial position, thus completing the task.

4.2 Holonomic constraint experiment

For simplicity's sake, the surface used to test the validity of the proposed approach is a sphere described by

$$\varphi_{\mathrm{v}}(\boldsymbol{x}_{\mathrm{v}}) = (x_{\mathrm{v}} - h)^2 + (y_{\mathrm{v}} - k)^2 + (z_{\mathrm{v}} - l)^2 - r^2 = 0, \tag{84}$$

where $\boldsymbol{x}_{\mathrm{v}} = \begin{bmatrix} x_{\mathrm{v}} & y_{\mathrm{v}} & z_{\mathrm{v}} \end{bmatrix}^{\mathrm{T}}$ is the vector that stands for the virtual environment task-space coordinates, $r = 0.1$ [m] is the radius, and $(h, k, l) = (0.4, 0, 0)$ [m] are the sphere center coordinates. It is important to note that, in contrast with other works, we added a third dimension $\mathbf{z}_{\mathrm{v}}$ in order to heighten the realism of the virtual reality application. For example, in [44, 49], the authors consider only two dimensions to test different control schemes for a haptic and a teleoperation system, respectively. The gains of the master manipulator, described in (63), and the virtual manipulator, are shown in **Table 1**.

Finally, by using the Generic Penalty Method, the surface parameters are $\alpha_{\mathrm{v}} = 0.002$, $\xi = 100$ and $\omega_{\mathrm{n}} = 200$.

4.3 Non-holonomic constraint experiment

As mentioned before, we cannot express a deformable surface implicitly, even when the operator perceives it as a sphere both visually and haptically. We use a

Variable (control law)	Value	Variable (virtual manipulator)	Value
$\boldsymbol{K}_{\mathrm{am}}$	0.0550 $\boldsymbol{I}$	$\boldsymbol{K}_{\mathrm{av}}$	0.20 $\boldsymbol{I}$
$\boldsymbol{K}_{\mathrm{pm}}$	0.0055 $\boldsymbol{I}$	$\boldsymbol{K}_{\mathrm{pv}}$	diag{0.2, 0.2, 0.2, 0.1, 0.1}
$\boldsymbol{K}_{\mathrm{fm}}$	10.050 $\boldsymbol{I}$	$\boldsymbol{K}_{\mathrm{fv}}$	0.20 $\boldsymbol{I}$
$\boldsymbol{\Lambda}_{\mathrm{xm}}$	0.2500 $\boldsymbol{I}$	$\boldsymbol{\Lambda}_{\mathrm{xv}}$	20.0 $\boldsymbol{I}$
$\boldsymbol{K}_{\beta\mathrm{m}}$	0.0100 $\boldsymbol{I}$	$\boldsymbol{K}_{\beta\mathrm{v}}$	1.00 $\boldsymbol{I}$
$\boldsymbol{K}_{\gamma\mathrm{m}}$	0.0150 $\boldsymbol{I}$	$\boldsymbol{K}_{\gamma\mathrm{v}}$	0.20 $\boldsymbol{I}$

***I** is the identity matrix, which has the appropriate dimensions.*

Table 1.
Gains from the control law and the virtual manipulator (Holonomic constraint experiment).

discrete representation similar to that presented in [52], where we assume the surface to comprise many neighboring planes defined by shared nodes. We propose a technique that comprises iteratively choosing a small neighborhood of planes where the contact will occur, depending on the position of the virtual tool. Subsequently, we associate the Lagrange multipliers in Eq. (70) with a pair of planes using the impulse-based technique for multiple rigid body simulations, [55]. The micro-collisions with this technique occur only in the chosen vicinity of the sphere and Lagrange multipliers replace the impulses preventing body interpenetration. In the collision's case detection algorithm, the implicit surface representation replaces the convex polyhedra decomposition, [56], using the Eq. (84). We did this for ease and to reduce the computational cost of the application, otherwise, the control algorithm sample time would increase. Considering the case where the needle is inside the sphere, but it may not move laterally. However, it may pivot to change orientation.

We adequately describe this kind of scenario by employing non-holonomic constraints. As mentioned in Section 1, non-holonomic constraints have been little exploited to represent the interaction with penetrable surfaces. For example, in [57] it is claimed that to model a surgeon's scalpel both holonomic and non-holonomic constraints could be employed by limiting the depth of its incision and the direction of its motion, respectively. However, there is not any analysis or modeling of this process in such work.

The most representative proposal is the derivation of the non-holonomic generalized unicycle model presented in [58], where a coordinate-free representation is used to model the insertion of a flexible needle into soft tissue. We employed a similar approach by using the homogeneous matrix representation, but taking into consideration both the kinematic and the dynamic model of the virtual robot and the fact that non-holonomic constraints are more intuitively got if we define them in task space coordinates. The computation of the Lagrangian multipliers for non-holonomic constraints, which is proposed in (71), is an important improvement regarding the cited works. The experiment comprised five degrees of freedom virtual manipulator interacting with a deformable sphere. Once in contact, the end-effector may not move laterally, i.e., along the ${}^{0}\boldsymbol{y}_{5\mathrm{v}}$ and ${}^{0}\boldsymbol{z}_{5\mathrm{v}}$ axes, after a conventional Denavit-Hartenberg allocation, but it may move along the ${}^{0}\boldsymbol{x}_{5\mathrm{v}}$ axis, i.e., along the pointing direction of the end-effector. The end-effector may rotate (pivoting) to change direction (as a three-dimensional version of the non-holonomic unicycle) and the Pfaffian matrix is computed as

$$\boldsymbol{A}_{\mathrm{v}}(\boldsymbol{x}_{\mathrm{v}})\boldsymbol{v}_{\mathrm{v}} = \begin{bmatrix} {}^{0}\boldsymbol{y}_{5\mathrm{v}} & \mathbf{0}_{1\times 3} \\ {}^{0}\boldsymbol{z}_{5\mathrm{v}} & \mathbf{0}_{1\times 3} \end{bmatrix} \begin{bmatrix} \dot{\boldsymbol{\rho}}_{\mathrm{v}} \\ \boldsymbol{\omega}_{\mathrm{v}} \end{bmatrix} = \mathbf{0} \tag{85}$$

The experiment has four steps, as shown in **Figure 17**. First, the virtual robot is in free motion and only the teleoperation part of the controller (77) is active. In the second part, we insert the needle into the sphere. Next, in the third part, approximately 45 degrees rotate the needle without changing its position. Finally, we insert the needle deeper into the sphere with a new orientation.

The force of the human operator in the lateral directions of the needle is difficult to measure directly with the force sensor. Instead, we take advantage of the projection of such forces on the torque of the master manipulator joint, we calculate $\boldsymbol{\lambda}_{\mathrm{m}}$ from (20), (23), and (73).

The gains of the master manipulator, described in (76), and the virtual manipulator, are shown in **Table 2**.

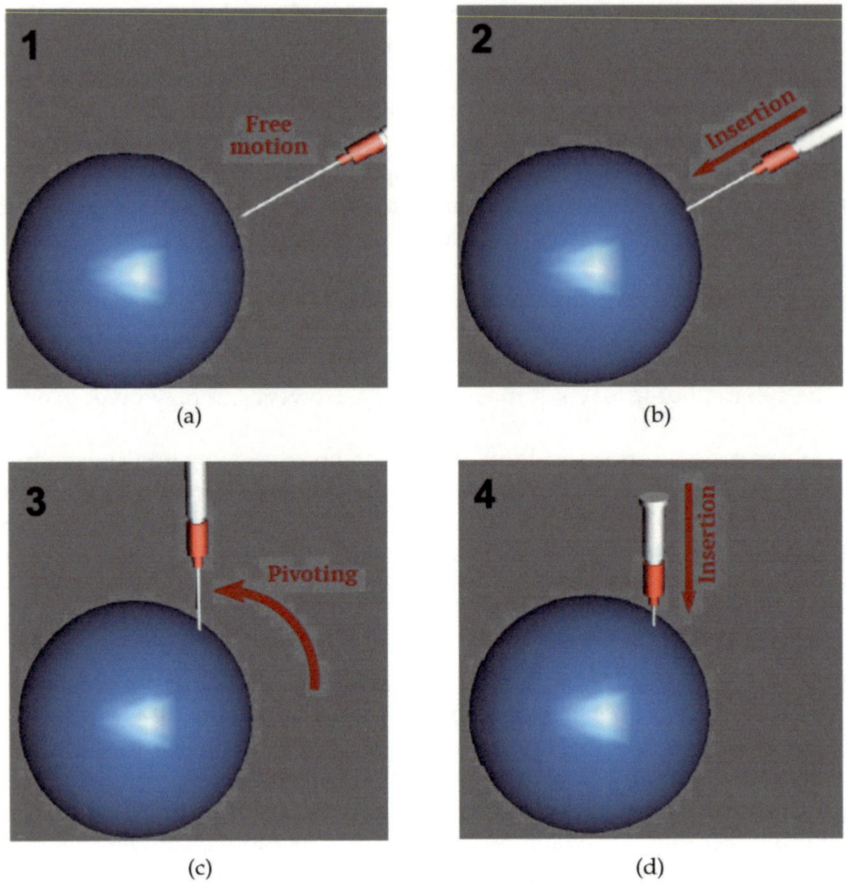

Figure 17.
Interaction sequence between the avatar and the non-holonomic virtual surface. (a) Free motion. (b) Insertion. (c) Pivoting. (d) Insertion.

Variable (control law)	Value	Variable (virtual manipulator)	Value
$\boldsymbol{K}_{\mathrm{am}}$	0.0550 $\boldsymbol{I}$	$\boldsymbol{K}_{\mathrm{av}}$	0.20 $\boldsymbol{I}$
$\boldsymbol{K}_{\mathrm{pm}}$	0.0550 $\boldsymbol{I}$	$\boldsymbol{K}_{\mathrm{pv}}$	diag$\{0.2, 0.2, 0.2, 0.1, 0.1\}$
$\boldsymbol{K}_{\mathrm{fm}}$	0.010 $\boldsymbol{I}$	$\boldsymbol{K}_{\mathrm{fv}}$	2.00 $\boldsymbol{I}$
$\boldsymbol{\Lambda}_{\mathrm{xm}}$	0.2500 $\boldsymbol{I}$	$\boldsymbol{\Lambda}_{\mathrm{xv}}$	20.0 $\boldsymbol{I}$
$\boldsymbol{K}_{\beta\mathrm{m}}$	0.0150 $\boldsymbol{I}$	$\boldsymbol{K}_{\beta\mathrm{v}}$	1.00 $\boldsymbol{I}$
$\boldsymbol{K}_{\gamma\mathrm{m}}$	0.0150 $\boldsymbol{I}$	$\boldsymbol{K}_{\gamma\mathrm{v}}$	0.20 $\boldsymbol{I}$

$\boldsymbol{I}$ is the identity matrix, which has the appropriate dimensions.

Table 2.
Gains from the control law and the virtual manipulator (Non-holonomic constraint experiment).

5. Conclusions

We present a proposal on chaptic interaction with holonomic and non-holonomic virtual constraints. Since extensive research on haptic interaction with rigid surfaces has been presented in the literature, the principal aim was to reproduce the forces generated by the interaction with soft surfaces from a force feedback approach.

Throughout the document, we introduced the theory to establish an optimal relationship between the visual and haptic interaction for the virtual reality system developed. The key lies in adapting the mathematical properties of the holonomic and non-holonomic constraints to the tool's kinematics in contact with a nonpenetrable and penetrable virtual surface, respectively. However, it is important to note that we made this adaptation to achieve haptic feedback purposes only and consider the basic contact properties of simulated rigid and soft tissues.

Adapting a teleoperation control scheme to a virtual reality system was the strategy to follow since it allowed embedding a robot's dynamic model into the virtual environment. By doing this, we addressed the teleoperated slave system as a problem of a virtual robot in constrained motion and either holonomic or non-holonomic constraints gave whose contact force.

We studied the differences between one or another representation, both mathematically and intuitively, and the particularities of each one. Among them, there is the fact that they could render forces using the *Generic Penalty Method* or the *Pfaffian constraints matrix*, respectively. We centered the interest on reproducing the contact with a soft surface by employing non-holonomic constraints.

The principal use detected is that such a method can render similar forces to those arising from contact between a tool and a penetrable surface. We can see, non-holonomic constraints have been used to reproduce the operator's tactile sensations with practical meaning. We can eventually use this approach in virtual reality medical simulators, and it presented the fundamentals to do that throughout this document. However, adapting the developed method to complex virtual environments, such as those found in the medical field, requires more research both in control and computer graphics. In the first case, adapting more accurately the teleoperation controller presented is necessary and makes it fit with the current methods of graphic computing. Finding more optimal ways to model force using non-holonomic constraints is essential to heighten the realism of the applications. Regarding graphic computing, it is necessary to design numerical methods that adapt more efficiently to the control algorithms designed and that are capable of running with continuum mechanics models.

Naturally, this will increase computational processing and require more analysis to establish the compensation between real-time processing and control performance, without sacrificing application realism. All this requires a wide range of knowledge that does not belong to control or computer graphics.

Acknowledgements

The authors are grateful for the support provided by PRODEP-BUAP ID90934 and the DGAPA-UNAM under grant IN117820.

Appendix

Algorithm 1: Deformable sphere

Require: Number of Parallels p.
Require: Number of Meridians m
Require: Radius r
Require: Offset r_{off}
Define: Vertex v_{ax}, v_{ay}, v_{az}
Define: Vertex v_{bx}, v_{by}, v_{bz}
Define: Vertex v_{cx}, v_{cy}, v_{cz}
Evaluate: $\Delta_1 = 180^\circ/p$ and $\Delta_2 = 360^\circ/m$

---

for all $i = 0$ to $p/2$ **do**
for all $j = 0$ to m **do**
$\alpha = i \times \Delta_1$
$\beta = j \times \Delta_2$
Calculate the vertices of the upper hemisphere using (81)-(83).
if avatar touches a superior plane **then**
Move the contact plane through the auxiliary normal force
Store data of the contact plane (based on p and m)
end if
Compute vertices of the lower hemisphere
if avatar touches an inferior plane **then**
Move the contact plane using (81)-(83) and the negative numbers on the z_v axis.
Store data of the contact plane (based on p and m)
end if
end for
end for
if avatar touches a plane **then**
Reordering adjacent planes
end if
for all $i = 0$ to $p/2$ **do**
for all $j = 0$ to m **do**
Draw plane
end for
end for

Author details

Pablo Sánchez-Sánchez[1*†], José Daniel Castro-Díaz[2†], Alejandro Gutiérrez-Giles[3†] and Javier Pliego-Jiménez[4†]

1 Facultad de Ciencias de la Electrónica, Benemérita Universidad Autónoma de Puebla, Puebla, Mexico

2 Facultad de Estudios Superiores ARAGON, UNAM, Mexico City, Mexico

3 Mechatronics Section, Electrical Engineering Department, CINVESTAV-IPN, Mexico City, Mexico

4 Applied Physics Division, Electronics and Telecommunications Department, CISESE-CONACYT, Ensenada, Mexico

*Address all correspondence to: pablo.sanchez@correo.buap.mx

† These authors contributed equally.

References

[1] Avgousti S, Christofouru EG, Panaydes AS, Voskarides S, Novales C, Nouaille L, et al. Medical telerobotic systems: Current status and future trends. Biomedical Engineering Online. 2016;**15**(96):1-44. DOI: 10.1186/s12938-016-0217-7

[2] Ballantyne GH, Moll F. The da vinci telerobotic surgical system: The virtual operative field and telepresence surgery. The Surgical Clinics of North America. Dec 2003;**6**(83):1293-1304. DOI: 10.1016/S0039-6109(03)00164-6. PMID: 14712866

[3] Basdogan C, De S, Kim J, Muniyandi M, Kim H, Srinivasan MA. Haptics in minimally invasive surgical simulation and training. IEEE Computer Graphics and Applications. 2004;**24**(2):56-64

[4] Hannaford B, Rosen J, Friedman DW, King H, Roan P, Cheng L, et al. Raven-ii: An open platformfor surgical robotics research. IEEE Transactions on Biomedical Engineering. 2013;**60**(4): 954-959

[5] Dy M-C, Tagawa K, Hiromi TT, Masaru K. Collision detection and modeling of rigid and deformable objects in laparoscopic simulator. In: Proc. SPIE 9415, Medical Imaging 2015: Image-Guided Procedures, Robotic Interventions, and Modeling, 941525 1-6 (18 March 2015). Orlando, Florida, United States; 2015. DOI: 10.1117/12.2081344

[6] Goertz, RC, Thompson WM. Electronically Controlled Manipulator. Nucleonics (US) Ceased Publication. 1954;**12**(11). DOI: 10.1146/annurev.ns.04.120154.002153

[7] Hollis RL, Salcudean S, Abraham DW. Toward a tele-nanorobotic manipulation system with atomic scale force feedback and motion resolution. In: IEEE Proceedings on Micro Electro Mechanical Systems, An Investigation of Micro Structures, Sensors, Actuators, Machines and Robots. Napa Valley, CA, USA; 1990. pp. 115-119. DOI: 10.1109/MEMSYS.1990.110261

[8] Khatib O, Yeh X, Brantner G, Soe B, Kim B, Ganguly S, et al. Ocean one: A robotic avatar for oceanic discovery. IEEE Robotics Automation Magazine. 2016;**23**(4):20-29

[9] Kim K, Song H, Suh J, Lee J. A novel surgical manipulator with workspace conversion ability for telesurgery. IEEE/ASME Transactions on Mechatronics. 2013;**18**(1):200-211

[10] Hansen T et al. Implementing force-feedback in a telesurgery environment, using parameter estimation. In: 2012 IEEE International Conference on Control Applications. Dubrovnik, Croatia; 2012. pp. 859-864. DOI: 10.1109/CCA.2012.6402708

[11] Sutherland IE. The ultimate display. In: Proceedings of the IFIP Congress. Vol. 2. London: Macmillan and Co; 1965. pp. 506-509

[12] Brooks FP, Ouh-Young M, Batter JJ, Jerome Kilpatrick P. Project gropehaptic displays for scientific visualization. SIGGRAPH Computer Graphics. 1990; **24**(4):177-185

[13] Chu CP, Dani TH, Gadh R. Multimodal interface for a virtual reality based computer aided design system. In: Proceedings of International Conference on Robotics and Automation. Vol. 2. Albuquerque, NM, USA; 1997. pp. 1329-1334. DOI: 10.1109/ROBOT.1997.614321

[14] Penn P, Petrie H, Colwell C, Kornbrot D, Furner S, Hardwick A. The haptic perception of texture in virtual environments: an investigation with two devices. In: Brewster S, Murray-Smith R, editors. Haptic Human-Computer

Interaction. Haptic HCI 2000, Lecture Notes in Computer Science. Vol. 2058. Berlin, Heidelberg: Springer; 2001. DOI: 10.1007/3-540-44589-7_3

[15] Hamza-Lup FG, Bogdan CM, Popovici DM, Costea OD. A survey of visuo-haptic simulation in surgical training. In: Proceedings on Mobile, Hybrid, and On-line Learning. Athens, Greece; 2011. pp. 57-62. ISBN: 978-1-61208-689-7

[16] Massie T, Salisbury JK. The PHANTOM haptic interface: A device for probing virtual objects. In: American Society of Mechanical Engineers Symposium on Haptic Interfaces for Virtual Environment and Teleoperator Systems. Vol. 1. Chicago, IL: Semantic Scholar; 1994. pp. 295-301

[17] Grange S, Conti F, Helmer P, Rouiller P, Baur C. Overview of the delta haptic device. In: Eurohaptics '01. Vol. 1. Birminghan, England: Semantic Scholar; 2001

[18] Conti F, Khatib O. A new actuation approach for haptic interface design. The International Journal of Robotics Research. 2009;**28**(6):834-848

[19] Finch M, Chi VL, Taylor RM, Falvo M, Washburn S, Superfine R. Surface modification tools in a virtual environment interface to a scanning probe microscope. In: Proc. Symposium on Interactive 3D Graphics, 13–24 New York, NY, USA, California, USA: Monterey; 1995

[20] Bliss JP, Tidwell PD, Guest MA. The effectiveness of virtual reality for administering spatial navigation training to firefighters. Presence: Teleoperators and Virtual Environments. 1997;**6**(1):73-86

[21] Rizzo A et al. Virtual environment applications in clinical neuropsychology. In: Proceedings IEEE Virtual Reality 2000 (Cat. No.00CB37048). New Brunswick, NJ, USA; 2000. pp. 63-70. DOI: 10.1109/VR.2000.840364

[22] Ruspini DC, Kolarov K, Khatib O. The haptic display of complex graphical environments. In: SIGGRAPH 97 Computer Graphics Proceedings Anual Conference. Los Angeles, California; 1997. pp. 140-147

[23] Salisbury K, Conti F, Barbagli F. Haptic rendering: Introductory concepts. IEEE Computer Graphics and Applications. 2004;**14**:24-32

[24] Escobar-Castillejos D, Noguez J, Neri L, Magana A, Benes B. A review of simulators with haptic devices for medical training. Journal of Medical Systems. 2016;**104**(40):177-185

[25] Duriez C, Dubois F, Kheddar A, Andriot C. Realistic haptic rendering of interacting deformable objects in virtual environments. IEEE Transactions on Visualization and Computer Graphics. 2006;**12**(1):1-12

[26] Terzopoulus D, Platt J, Barr A, Fleischer K. Elastically deformable models. Computer Graphics. 1987;**21**(4):205-214

[27] Sclaroff S, Pentland A. Generalized implicit functions for computer graphics. SIGGRAPH Computer Graphics. 1991;**25**(4):247-250

[28] Minsky M, Ming O, Steele O, Brooks FP, Behensky M. Feeling and seeing: Issues in force display. SIGGRAPH Computer Graphics. 1990;**24**(2):235-241

[29] Adams RJ, Hannaford B. Stable haptic interaction with virtual environments. IEEE Trans. on Robotics and Automation. 1999;**15**(3):465-474

[30] Bayo E, Avello A. Singularity-free augmented lagrangian algorithms for constrained multibody dynamics. Nonlinear Dynamics. 1994;**5**(2):209-231

[31] Zilles CB, Salisbury JK. A constraint-based god-object method for haptic display. In: Proceedings 1995 IEEE/RSJ International Conference on Intelligent Robots and Systems. Human Robot Interaction and Cooperative Robots. Vol. 3. Pittsburgh, PA, USA; 1995. pp. 146-151. DOI: 10.1109/IROS.1995.525876

[32] Mavhash M, Hayward V. High-fidelity haptic synthesis of contact with deformable bodies. IEEE Computer Graphics and Applications. 2004;**24**(2): 48-55

[33] Castro-Díaz JD, Sánchez-Sánchez P, Gutiérrez-Giles A, Arteaga-Pérez MA, Pliego-Jiménez J. Experimental results for haptic interaction with virtual holonomic and nonholonomic constraints. IEEE Access. 2020;**8**: 120959-120973

[34] Monroy C, Kelly R, Arteaga M, Bugarin E. Remote visual Servoing of a robot manipulator via Internet2. Journal of Intelligent and Robotic Systems. 2007;**49**:171-187

[35] Rodríguez A, Basañez L, Colgate JE, Faulring EL. Haptic display of dynamic systems subject to holonomic constraints. In: IEEE Int. Conference on Intelligent Robots and Systems. France: Nice; 2008

[36] Montagnat J, Delignette H, Ayache N. A review of deformable surfaces: Topology, geometry and deformation. Image and Vision Computing. 2001;**19**: 1023-1040

[37] Heredia SA, Harada K, Padilla-Castaneda M, Marques-Marinho M, Márquez-Flores JA, Mitsuishi M. Virtual reality simulation of robotic transsphenoidal brain tumor resection: Evaluating dynamic motion scaling in a master-slave system. The International Journal of Medical Robotics and Computer Assisted Surgery. 2019;**15**(1): 1-48. DOI: 10.1002/rcs.1953. Epub 2018 Oct 18. PMID: 30117272; PMCID: PMC658796

[38] Maurel W, Wu Y, Magnenat N, Thalman D. Biomedichal Models for Soft Tissue Simulation. Heidelberg, Germany: Springer; 1998

[39] Faure F, DuriezHervé C, Delingette H, Allard J, Gilles B, Marchesseau S, et al. SOFA: A multi-model framework for interactive physical simulation. In: Payan Y, editor. Soft Tissue Biomechanical Modeling for Computer Assisted Surgery. Studies in Mechanobiology, Tissue Engineering and Biomaterials. Vol. 11. Berlin, Heidelberg: Springer; 2012. DOI: 10.1007/8415_2012_125

[40] Selig JM. Geometric Fundamentals of Robotics. New York, USA: Springer Science and Business; 1996

[41] Xiaoping Y, Sarkar N. Unified formulation of robotics systems with holonomic and nonholonomic constraints. IEEE Transaction on Robotics and Automation. 1998;**14**:640-650

[42] Murray RM, Li Z, Sastry SS. A Mathematical Introduction to Robotic Manipulation. Boca Raton, Florida, USA: CRC Press; 1994

[43] Luca AD, Oriolo G. Modelling and control of nonholonomic mechanical systems. In: Angeles J, Kecskeméthy A, editors. Kinematics and Dynamics of Multi-Body Systems. CISM International Centre for Mechanical Sciences (Courses and Lectures). Vol. 360. Vienna: Springer; 1995. DOI: 10.1007/978-3-7091-4362-9_7

[44] Faurling EL, Lynch KM, Colgate JE, Peshkin MA. Haptic display of constrained dynamic systems via admittance displays. IEEE Transactions on Robotics. 2007;**23**:101-111

[45] Arteaga MA, Gutiérrez-Giles A, Pliego-Jiménez J. Bilateral teleoperation.

In: Local Stability and Ultimate Boundedness in the Control of Robot Manipulators, Lecture Notes in Electrical Engineering. Vol. 798. Cham: Springer; 2022. DOI: 10.1007/978-3-030-85980-0_9

[46] Garcia-Valdovinos LG, Parra-Vega V, Arteaga MA. Higher-order sliding mode impedance bilateral teleoperation with robust state estimation under constant unknown time delay. In: Proceedings, 2005 IEEE/ASME International Conference on Advanced Intelligent Mechatronics. Monterey, CA, USA. pp. 1293-1298. ISBN:0-7803-9047-4. DOI: 10.1109/AIM.2005.1511189

[47] Arteaga MA. Tracking control of flexible robot arms with a nonlinear observer. Automatica. 2000;**36**(9):1329-1337

[48] Arteaga MA, Castillo-Sánchez A, Parra-Vega V. Cartesian control of robots without dynamic model and observer design. Automatica. 2006;**42**(3):473-480

[49] Gutiérrez-Giles A, Arteaga-Pérez MA. Transparent bilateral teleoperation interacting with unknown remote surfaces with a force/velocity observer design. International Journal of Control. 2019;**92**(4):840-857. DOI: 10.1080/00207179.2017.1371338

[50] Gudiño Lau J, Arteaga MA. Dynamic model and simulation of cooperative robots: A case study. Robotica. 2005;**23**:615-624

[51] Shreiner D, Sellers G, Kessenich J, Licea-Kane B. OpenGL Programming Guide Eight Edition, The Official Guide to Learning OpenGL. The Khronos OpenGL ARB Working Group. Ann Arbor, Michigan: Addison-Weasley; 2013. ISBN 978-0-321-77303-6

[52] Zafer N, Yilmaz S. Nonlinear viscoelastic contact and deformation of freeform virtual surfaces. Advanced Robotics. 2016;**30**(4):246-257

[53] Barbagli F, Salisbury K. The effect of sensor/actuator asymmetries in haptic interfaces. In: 11th Symposium on Haptic Interfaces for Virtual Environment and Teleoperator Systems, 2003. HAPTICS 2003. Proceedings. Los Angeles, CA, USA; 2003. pp. 140-147. ISBN:0-7695-1890-7. DOI: 10.1109/HAPTIC.2003.1191258

[54] Yang C, Xie Y, Liu S, Sun D. Force modeling, identification, and feedback control of robot-assisted needle insertion: A survey of the literature. Sensors (Basel). 12 Feb 2018;**18**(2):1-48. DOI: 10.3390/s18020561. PMID: 29439539; PMCID: PMC5855056

[55] Constantinescu D, Salcudean SE, Croft EA. Haptic rendering of rigid contacts using impulsive and penalty forces. IEEE Transactions on Robotics. 2005;**21**(3):309-323

[56] Kim YJ, Otaduy MA, Lin MC, Manocha D. Six-degree-of-freedom haptic display using localized contact computations. In: Proceedings 10th Symposium on Haptic Interfaces for Virtual Environment and Teleoperator Systems. HAPTICS. Vol. 2002. Orlando, FL, USA; 2002. pp. 209-216. ISBN: 0-7695-1489-8. DOI: 10.1109/HAPTIC.2002.998960

[57] Faurling EL, Lynch KM, Colgate JE, Peshkin MA. Haptic interaction with constrained dynamic systems. In: Proceedings of the 2005 IEEE International Conference on Robotics and Automation. Barcelona, Spain; 2005. pp. 2458-2464. ISSN: 1050-4729. DOI: 10.1109/ROBOT.2005.1570481

[58] Webster RJ III, Seob J, Cowan NJ, Chirikjian GS, Okamura AM. Nonholonomic modeling of needle steering. The International Journal of Robotic Research. 2006;**25**(5–6):509-525. DOI: 10.1177/0278364906065388

Section 3

Haptic for Digital Heritage

Chapter 5

How to Use Haptic Technology in Interactive Digital Documentation of Heritage

Vladimír Hain, Roman Hajtmanek and Dušan Kočlík

Abstract

Virtual restoration of the extinct heritage is a method of reconstruction of an already destroyed work in a virtual environment. It represents a way and an opportunity to reenter the remodeled simulated space interactively and experience its contemporary atmosphere and former author expression. In addition to visual and acoustic experiences, haptic technologies represent the potential for expanding sensory perception, which is not yet sufficiently used in the architectural sector. This study focuses on defunct and endangered works of interior architecture and industrial heritage, which were significant at the time of their inception and shaped the direction of the industry. Especially in the case of cultural and spiritual heritage, we focus on interiors, which, by their short-term nature, are neither objective nor physically documentable. Selected extinct works for which there was enough data or there was still the possibility of consultation with a living author were experimentally virtually reconstructed. Using haptic technologies, we have expanded the observer's ability to interactively analyze space and its context through User Tracking of observers. The data obtained in this way continue to help the creators of the architecture set new starting points and limits for the current creation and design as well.

Keywords: architecture, heritage, digital documentation, haptic technology, user tracking

1. Introduction

Current digital technologies allow us to design, document, preserve, evaluate, and popularize cultural heritage and architectural heritage in many ways. However, their potential is not always fully exploited [1]. The chapter is aimed to explore form-forming factors of architectural space and theoretical research, between monument preservation, Haptic Technology, and architectural practice.

The study represents identified factors that affect the efficiency and quality of design process in the cooperation with modern technologies, documentation, and conservation process of heritage sites. It deals with the opportunities of transfer research results from the futuristic disciplines as well. In this case, the paper examines the study "Reconstruction of old industrial Power plant in Piestany" and describes one of the possible solutions based on the mixed reality (MR) application. The opportunity to experience this kind of an industrial object with multiple senses (sight, hearing, smell, touch) in MR delivered a unique personalized haptic experience and immersive memories about lost heritage.

Developed presentations, mixed reality interactive models nowadays can create attractive interpretation of this rich source of experiences and knowledge. The interdisciplinary research team at the Slovak University of Technology in Bratislava Faculty of Architecture and Design focuses systematically their work on applications of virtual reality (VR) by merging different sensorial inputs from mixed reality and real environment. The article is focused to explore opportunities for incorporation of haptic technologies into monument preservation, research of virtual and mixed reality and architectural practice [2].

2. Theoretical scope

Haptic technology has excellent potential to help society in their daily lives, design, or education. In schools or museums of technology around the world, there are innovative creations of the human spirit, which are often presented in a way that is in comparison to other media less attractive for the contemporary audience. Therefore, the contemporary trend is the development of interactive kind of the presentation of physics laws and technology. These types are capable of making technology museums more inspiring and of enabling the interactive use of this plentiful source of knowledge. Too many historical buildings were destroyed, and they no longer exist, but historical archive documents, drawings, or photographs have been preserved. Some buildings remain in the living memory, or few physical fragments have been preserved. This technical documents and protected parts of the building may propose data for a digital presentation of the significant design or industrial monument. The interpretation of a hypothetical reconstruction by mixed reality can serve to better understand the culture, history, and technology by the public [3]. The virtual presentation of the model can serve as a haptic presentation of the extinct design, technical and cultural heritage as well.

Haptic technologies have been explored in virtual arts, such as sound synthesis or graphic design and animation [4]. The potential of their use is in the whole breadth of virtuality continuum (**Figure 1**). For the ordinary presentations in practice is used mainly augmented reality (AR) and virtual reality (VR) of displayed types of realities differ according to degree of reality.

The taxonomy of Milgram and Kishino provides a way of contrasting different types of mixed reality. This paper is focused on different fusions of other various sensorial inputs from real human life as smell, touch, and hearing with virtual or mixed environment [5].

The theory of didactics confirms that the senses are for people portals of information. Some people learn by sight, hearing, or by certain kind of activity (**Figure 2**). Each of us prefers a different method and way of teaching. The use of the combinations of senses is typical for "mixing learning styles" [7].

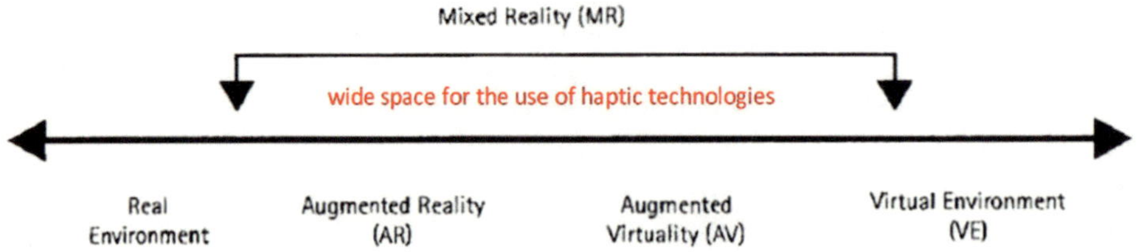

Figure 1.
Virtuality continuum diagram by Milgram and Kishino (Steed, 2013) [5].

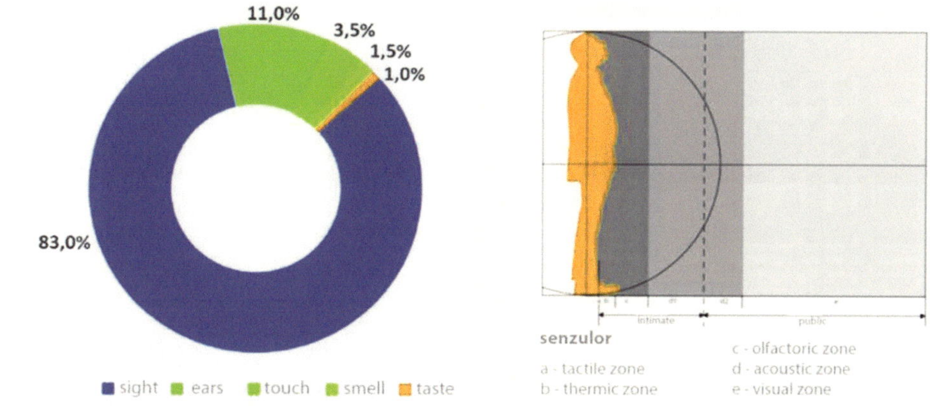

Figure 2.
Graph of sensory reception (M. Ganobjak, V. Hain, 2014) Picture of "Senzulor" was for the first time graphically illustrated by Prof. Robert Špaček in 1985. The term was created/used as a parallel by Modulor (authors: I. Keppl and R. Špaček, FAD STU, 1986) [6].

We receive a different percentage of information with every sense [6], and everyone remembers it differently. A difference needs to be created between receiving and remembering of the information. The most of the information we receive visually. By hearing, it is in comparison significantly less. We remember 20% of what we hear, 30% of what we see in visual form, and 90% of what we are actively doing [8].

Mixed reality actively uses mainly the first two human senses (sight and hearing) through which we receive the most of the information. Kinesthetic style of education uses activity of body and engages all senses (other three) without preference. It is proven that the best learning effectiveness is the way of learning through a combination of styles. Although the representation of other senses is negligible in receiving information, it appears that combinations of activating multiple senses are highly effective. This way, one can remember up to 80–90% of what one hears, sees, and does at once. It can be stated that the sensory overlap with which the information was captured creates stronger links between them for remembering. This is absent in the usual case of selective perceptions.

There are several cases of people with hearing, visual, or other disabilities that need to be kept in mind. In this case, one or more senses are missing, so they are replaced or compensated by another. Each situation is unique and different, it would be appropriate to pay special attention to each person with regard to their characteristics. However, it is not possible to set a specific tactile exposure for everyone. Universal design rules are offered as if they were the opposite of barrier-free design. It is a design for the widest possible range of users and not just for a narrowly specified group. Here it is important to create a quality exhibition that is inspiring and universal for everyone. One of the solutions to achieve such a balanced state is to create an exhibition and at the same time ensure that every single exhibit is perceived by several senses at the same time. This will provide the observer with fuller information. In addition, such an exposure to tactile or mixed reality allows a clearer situation to be understood and remembered not only by children but also by people with limited sensory abilities.

Such a prepared and focused presentation will bring visitors a new experience and allow them to perceive the laws of nature, often from a different perspective. The fun factor is also an integral and important part. It is usually a pleasant refreshment in the amount of informative information that comes to our attention.

The image of the Senzulor (**Figure 3**) shows the reach of our human senses. It shows the radius of the information we are able to receive in this sense. The eyes

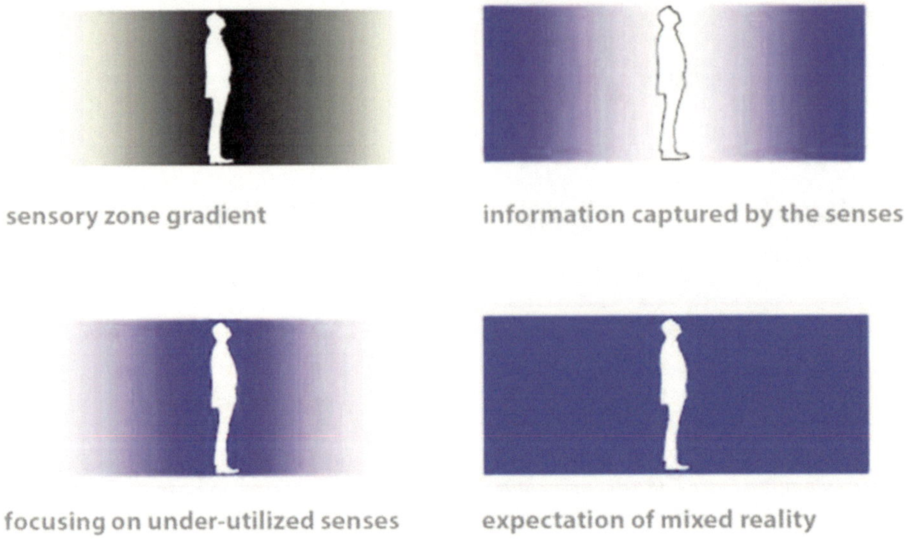

Figure 3.
Inverse sensory orientation of exposure. Combinations of sensory perception affect the overall impression (scheme: M. Ganobjak, V. Hain, 2014) [6].

capture a lot of information, but at the same time we are overwhelmed with visual information. Therefore, it is possible to use the method of inverse engagement of the senses. There are not many educational presentations that are tactile, haptic, acoustically olfactory or by taste.

Just as we perceive the stimulus closer to the body, it may leave a larger memory footprint. The human being subconsciously prefers those stimuli and impulses from the environment that act closer to the body surface. This proximity leads to an approximately defined sequence of its sensory zones from the tactile zone through the olfactory zone, the thermal zone, the acoustic zone to the human most dominant visual zone. Irritation of human receptors affects the perception of the environment, behavior, and orientation in space as well as the overall relationship to our environment. The center of gravity is activated by the sensory organs to determine the size and character of the individual frameworks of human zones. This dependence is expressed by the Senzulor.

All of our senses provide information about the properties of the external environment. Different organized and developed sensory organs with different sensitivities and complexities can only receive the same information as well as several pieces of information at the same time. Similar combinations of our sensory perceptions affect a person's overall impression, feeling, or condition in multiple situations. These phenomena are positively or negatively reflected especially in the perception of presentations, and therefore, it is important to pay close attention to them during designing mixed reality as well.

By involving multiple sensory stimuli, the information flow is enriched, making it easier to compare the user experience with a real experiential situation [9] that is closer to innate learning and thus to collect relevant data on user perceptions. Such data are mainly used as feedback, which could improve the future designs of other installations and exhibitions. There are many techniques for processing spatial and haptic information. The space can be sketched, 3D scanned or measured using classic techniques, and compared with suitable project documentation. Then it is necessary to model it accordingly in the form of a virtual 3D model. The individual characteristic surfaces need to be arranged in order to create textures with suitable qualities such as texture, reflection, color, etc. For obscure or unpreserved surfaces, it is possible to use photographs or retouched techniques or replace them with equivalent textures from similar objects.

3. Methodology

Haptic didactic tools educate "with an emphasis on the active and creative learning, not just passive reception of information." Interactivity allows two-way communication, and the student thus has the opportunity to actively intervene in the operation through the user interface program and not just passively participate (receive) its content. This increases clarity, motivation, and desire of students to learn. The basic advantages of interactivity innovation are "activity pupil, increasing pupil attention, motivation and desire to learn, actively and creatively engage in educational activities, a positive attitude and interest in the curriculum, etc."; ultimately better semantic connections and understanding of the curriculum.

Interactivity and immersiveness are important keys on exploration of virtual reality game. There is an issue involving haptic as part of stimulating interactivity between virtual characters and players in order to obtain more attention from players [10]. Haptic interfaces can create combinations of mechanical signals that do not have counterparts in real environments [11]. This allows creating haptic virtual environ in which entirely new haptic sensory experiences are possible (**Figure 4**).

The main goal of this research was to discuss the basics of effective use of haptic virtual environments in research of applications involving user sensory testing. To illustrate this intention, this chapter also discusses some recent discoveries in haptic perception, in which haptic presentation has played an important role in digital documentation of heritage; in this case, study of an industrial heritage.

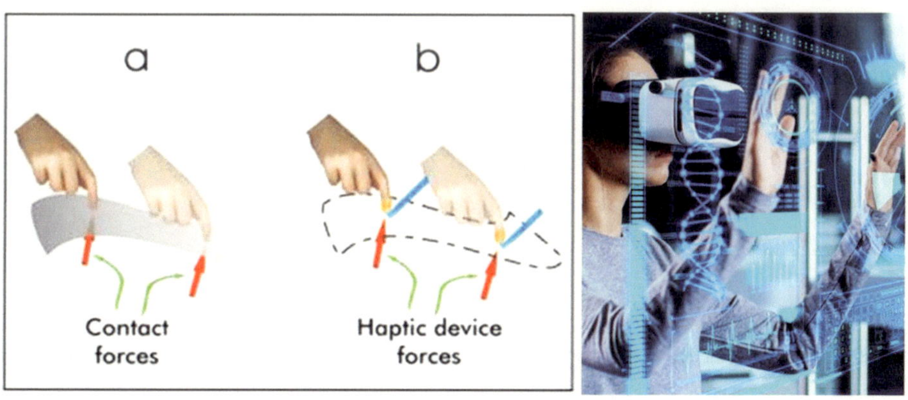

Figure 4.
(a) Haptic perception in everyday environments. (b) In contrast, haptic perception in virtual environments (scheme: [11]).

4. Case study

Digital documentation and presentation by haptic technology in the old power plant in the Piešťany city

The presented case study presented in this chapter is an example of the implementation of the methodology of the previous research chapter. It focuses on the use of virtual and mixed reality as an analytical tool for the design of exhibition space. In this way, a fuller exploration of new educational and simulation techniques in industrial spaces is ensured. The old power plant for heavy oil burning in Piešťany was built in 1906 as one of the first of its kind in the former Austro-Hungarian Empire. Later, the plant only provided distribution and energy transformation till the 1990s. The machinery hall originally had six diesel engines and generators. Now there is a multifunctional hall for scientific devices, exhibitions, and cultural events. Archival documents about the original state of the machinery hall allowed the exact appearance to be replicated through MR [12] (**Figure 5**).

After conversion, the building is now used as a technical science museum, which interactively educates about the energy and electricity sector (**Figure 6**). The building can currently be used for multifunctional common purposes, and at the same time, visitors can learn more additional information about the history of electricity in Slovakia. The exhibition is a hybrid of mixed reality, 3D haptic models, virtual reality, and physical industrial objects. These model solutions are defined according to the architectural value of the monuments [13]. The proportions, materials, and details for the 3D model were derived from preserved and functional historic diesel engines from the Technical Museum in Vienna. Photographic processes took 3 days through 3D scanning. Based on interdisciplinary cooperation and 3D animation of a historic engine MR exhibition was created.

The 3D model serves as a reference from which it was possible to analogously create proportions of details 1:1 (**Figures** 7 and **8**) and draw them in a new complete 3D model of the building. Based on measurements on-site and archival research, it was found how the building was originally built according to plan in 1906. Further historical research identified all periods of building extensions and various stages of building outlook (1920–1945). For the purposes of this case study, it was decided to visualize just the first and oldest period of 1906 [12].

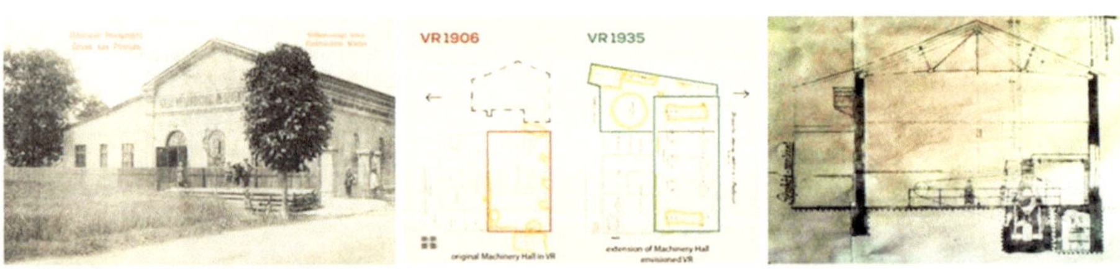

Figure 5.
Archival documents of the building from the National Archive in Trnava from 1906 to 1938 (photo: V. Hain, M. Ganobjak, 2010).

Figure 6.
Project of reconstruction of old power plant in Pieštany: M. Ganobjak, V. Hain, M. Paško, Z. Zacharová, 2014 (photo: P. Safko, 2014).

Figure 7.
Original diesel engine from Vienna Technical Museum compared with photogrammetry of 3D model via software Capture Reality and AGISoft (photo and 3D model: O. Virág, 2016).

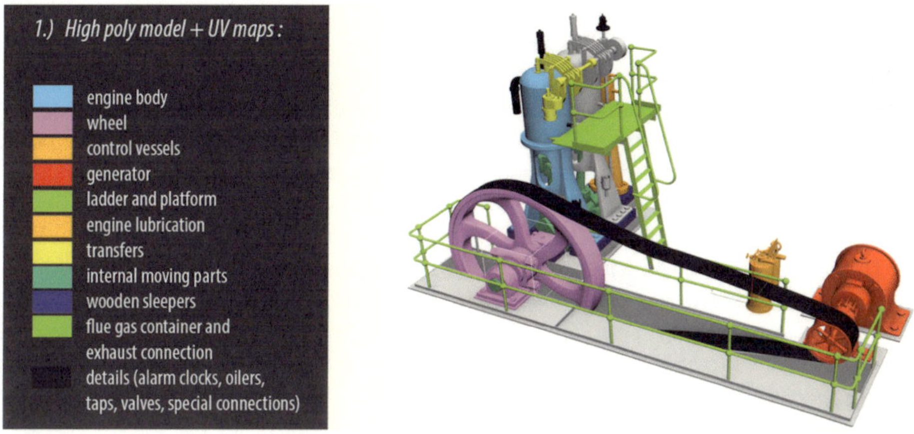

Figure 8.
Final VR 3D model of the virtual presentation was presented by VR headset Oculus Rift in the Power Plant Piešťany, (3D model: O. Virág, 2016).

The user can experience the atmosphere of a characteristic industrial space design in original realistic quality, along with real-time sounds and animations. VR objects and a 3D model were prepared in Unreal Engine 4, which provides photorealistic images with high-quality surfaces, textures, and lighting. The outputs are suitable for all these selected tested devices: HTC Vive, Oculus Rift, Cyberith [2].

The VR scene for the power plant created in 1906 (**Figures 8** and **9**) is intended for education and visual communication of technical information, but it also builds on the diversity of educational and multisensory exposition, which is more universal. The target group is students, all visitors to the practical science center of the EP, but also experts in the field of electrical engineering, whom the exhibition created in this way can entertain but mainly teach the general public.

3D model of the engine room seeks to eliminate the extreme situations of negative emotions of the space; it is "phobia free." MR respects the senses and aims to eliminate potential negative emotions. The space is becoming appropriate. MR and VR evoke feelings from original environment supplemented by authentic sounds and smell that invoke an industrial atmosphere. On the magic date of Friday, May 13, 2016, the virtual reality project was presented for the first time in the old power

Figure 9.
3D model of the original machine and mixed reality presentation with VR headset Oculus Rift in the power plant in Piešťany. For visitors it was possible to compare the current status and historical status—an overlay of physical and virtual reality (photo: O. Virág, V. Hain, Ľ. Dait, M. Ganobjak, 2016).

plant in Piešťany through Oculus Rift glasses for VR (Video 1, https://www.youtube.com/watch?v=Pk-8gCx03WM&feature=youtu.be).

The presentation is fully animated with the possibility of synchronized human movement in space. The exhibition is thus interactive and creates a subjective experience. The audiovisual design in the original old machine hall of the old power plant sensually complements it with the historical scent of black oil (unrefined diesel). This greatly affects the imagination of the observer, allowing him to be better immersed in the experience for long-term storage of sensory information. At the same time, the MR presentation premises is a more advantageous form for a wider audience of all ages and for people with certain forms of disability. It's a so-called as a "window to the past." This kind of mixed reality experience and visitors has proven that it is a suitable tool for commemorating the extinct heritage and reinterpreting its significance for the present (**Figure 10**).

The virtual Machinery Hall was tested by virtual tracking of the visitors. The mentioned motivation, inducing natural behavior, was taking photos of what they see. The reward system, which was linked to the real and also supported their natural behavior, was displaying their photographs and movements on the additional display. In addition, taking photos by the visitors marked the most interesting views and locations in the presented virtual space. Subsequently to the virtual

Figure 10.
MR application testing by students of the University of the Third Age of the FAD STU in Bratislava (photo: V. Hain, 3D model: O. Virág, 2016).

exploration of the space, the brief questionnaire was given to them. This questionnaire concerned their feelings in the virtual environment and the overall quality of the virtual presentation [14].

Similar presentations using VR is appropriate and could be also adapted to people with different disabilities—the virtual movement through space for people with movement disabilities, visual space for people with hearing disabilities, brightness and contrast color scheme for people with seeing impairments, and rich sound experience for the blind people.

The virtual presentation is not limited by the visual or graphical style, it could be hyper-realistic, sketchy, or abstract, and it is also saving space and is very customable. The currently unavailable spaces of the power plant in Piešťany are opened to public, and its past capabilities as circulation of fuel and cooling water through the past generators in the Machinery Hall are explained by the haptic diagrams. The authentic remaining equipment is complemented by educational presentation diagrams in various languages explaining its functioning by LCD touch panels.

On the wall and floor of the hall, the timeline with augmented reality presenting the electricity utilization is drawn. By focusing the tablets on the individual points of the timeline, the technology of the specific period is presented by the animation. The interactive installations are complemented by the Tesla coil, which is hanged on a steel rope above the heads of the observers, and it is throwing lighting above them. The turret room has also a stainless-steel ball in the middle, which is a Van der Graaf generator that bristles the hair of the visitors who are touching it (**Figure 11**).

The visitors reported that to move through virtual space without their avatar body was not comfortable experience. In the beginning of their virtual visits, they were a bit confused and disoriented, but in a short time, they adapted to that state and examined the space without obstructions. Use of the real environment as an anchor point for visitors' orientation and location in space showed to be very efficient for successful education, because the brain distinguishes the additionally given information in virtual reality, and it directly connects them with the real place. On the other hand, using mixed reality in this case study appeared to be a very practical tool for presentations at different places, outside of the original site of old power plant in Pieštany.

Here appeared the first hint and requirement of users for the implementation of additional haptic technologies, with which they would feel more anchored in space,

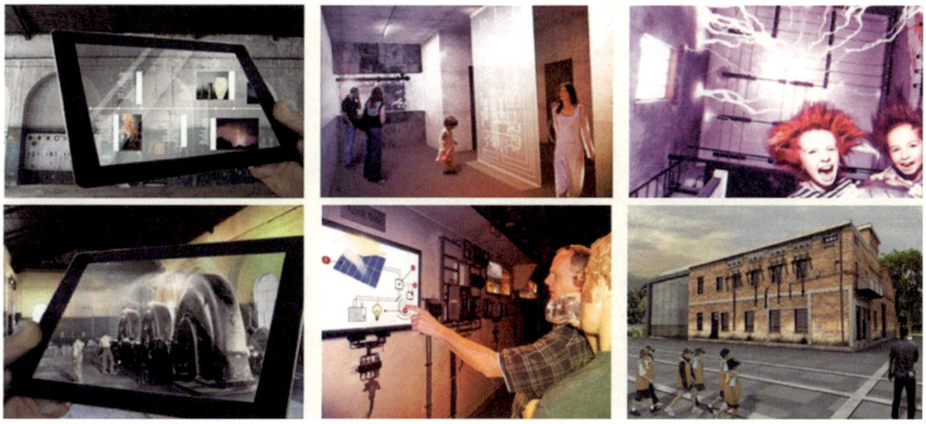

Figure 11.
Mixed reality exhibition in the old power plant in Pieštany with augmented reality, virtual reality, and of original engine equipment (design and photo: V. Hain, M. Ganobjak).

more confident in understanding what is safe and what is risky. Acoustic or vibration signals would be appreciated by most users.

5. Visitor tracking and virtual exhibition evaluation

The virtually reconstructed machinery hall of power plant in Piešťany was also presented at the Night of European Researchers in Bratislava. At this event, the tracking of the visitors in this virtual installation was included. To induce natural behavior in visitors, we motivated them by the ability of taking photos of the virtual machinery. The second screen displayed the taken photos and their motions as reward system even more supporting the motivation of visitors. The photographs taken by users marked the most attractive places and motives of the virtual exhibition.

When the visitors finished their virtual observation, they answered a brief form containing questions about the comfort of VR and quality of this type of presentation.

Motions and gazes of the visitors in VR were noted every 0.3 second. These data were gathered with positions from which the photographs were taken, into the dense cloud of points to process them in the subsequent research. Visitors' motions were also noted via the heat map by the contrast trace. When visitors spent more time on a specific position, the trace became more contrast. These data notation enabled to visualize the attractiveness of certain places and to process them by supervised machine learning to create a prototype of an analytical instrument for evaluation of similarly designed virtual exhibitions (**Figure 12**).

The prototype of the analytical tool for such an evaluation is a statistical model based on the artificial neural network (ANN) trained by supervised learning. By the supervised learning, the ANN is learning the relations and links between the pairs of related input and output samples [15].

To teach the ANN, the planar heat map with visitors' motions was resampled to 40 × 66 pixels and sampled in 0.6 m, which is the size of human module, usually used in architectural design. Sampling the heat map, divided it to samples, each with four pixels. These samples were positioned in the original grid of 40 by 66 positions. In these positions, the 3D model of the exhibition was processed by the isovist tool, which is quantifying the spatial openness and visibility by measuring the distances from the certain positions to their surrounding objects.

Figure 12.
Users' tracking data: left—heat map of tracked users' motions in plan, right—point cloud of tracked users' view locations and positions, blue points are photographed views (R. Hajtmanek, 2019).

In this case, 24 distances from every location in the grid to the surrounding objects were measured. The sums of each 24 distances quantified the openness and visibility of the space in every location of the square grid. This analysis of the space openness was also noted via the planar heat map, equally sampled into four-pixel samples as the heat map of the tracked visitors' motions. The measured objects in the exhibition were also categorized via its significance. Categorization of the objects was made of three groups according to their significance: 1—windows and walls, 2—subsidiary hall's equipment, 3—the most important and attractive diesel machines in the hall. Every distance measure contained then also the information of significance of the measured object, which was visible from the certain location in the grid.

5.1 Results of the evaluation

The supervised learning of the ANN contains training and testing phases. In the testing phase, AAN is trained on the training set, consisting 80% of the total samples count. After the training phase, it is tested in the testing phase on the remaining 20% of the samples. The comparison between the test and original data then validates the learning of the ANN.

Based on the learning, the AAN generated the new heat maps of visitors' motions, from the input data of spatial openness and objects' importance. These newly generated maps were then compared with the original tracked data of the visitors' motions. The original and generated heat maps were colored and blurred to highlight the similarities or differences (**Figure 13**).

Graphical comparison of the heat maps validated the ANN learning in the training phase, as these parts of the images are similar. Comparison of the image parts generated during the training phase also shows similarities but with some inaccuracies. Still, it is possible to declare that some relations were learned by ANN as the

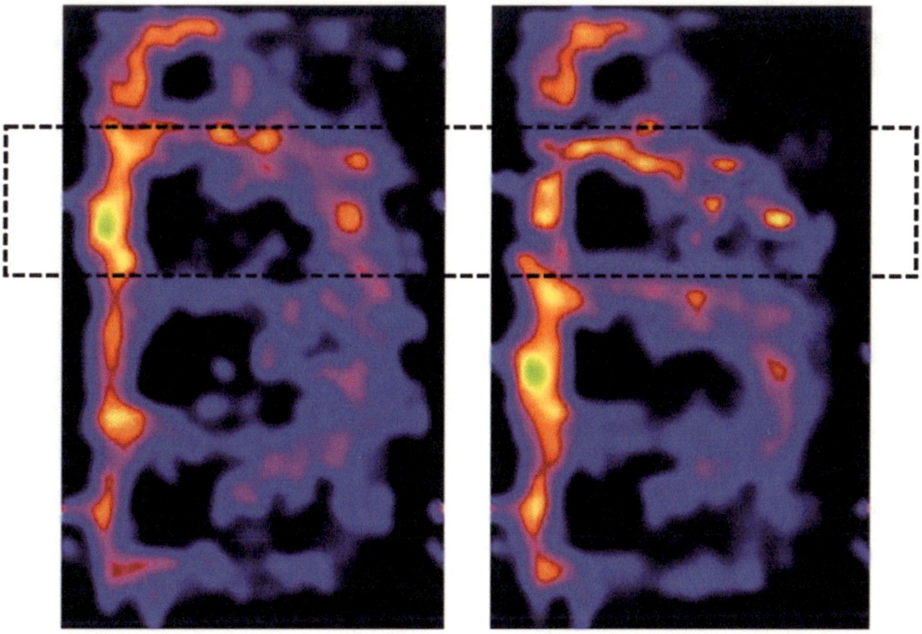

Figure 13.
Comparison of the original and generated maps of the visitors' motions left—original blurred and recolored heat map, right—ANN generated blurred and recolored heat map. Area marked by the dashed rectangle was generated in the test phase (R. Hajtmanek, 2019).

recognition of the attractive space between the machines and windows and motions around the objects. With these outcomes, the prototype of this tool based on the ANN validated that it is possible to evaluate similar designs of the virtual exhibitions by predicting statistical response from its future visitors. Such an evaluation during the design process can then bring more attractive and better suited further virtual presentations.

6. Screening

The precisely modeled 3D representation of Machinery Hall in power plant in Piešťany was very captivating to the general public, but also to the energy professionals from Západoslovenská energetika a.s.—electricity supplier in the west of Slovakia. Together with iPartner and Živica–Center for Environmental and Ethical Education, an educative and interactive quiz game for primary schools was developed (**Figure 14**). Team from Faculty of Architecture and Design STU created an interactive application based on VR game, through which pupils solve tasks related to the subjects of physics, chemistry but especially electric energy. In addition to the classic haptic game with cables and a plasma lamp, they could also try themselves education virtually by visiting the Piešťany Power Station in 1906 via the VR application.

This application has already been tested at the Pavol Horov Primary School in Devínská Nová Ves—Bratislava. This quiz was tried by children from 12 to 15 years of age using the VR headsets. The screening in the schools showed that this way of education increased the interest of pupils strongly.

Pedagogues without VR experiences were interested by implementation of similar interactive methods in their future teaching process.

Experts in the field of industrial heritage and its pedagogy see the significance of presentation by virtual 3D models of lost historic objects in a few points:

- These installations are presenting the site to the wider public, and they serve as a reminder of local history.

- They are opening to the public, but also professional discussion about the site and its future image.

- They are reimagining the ideas about present and future.

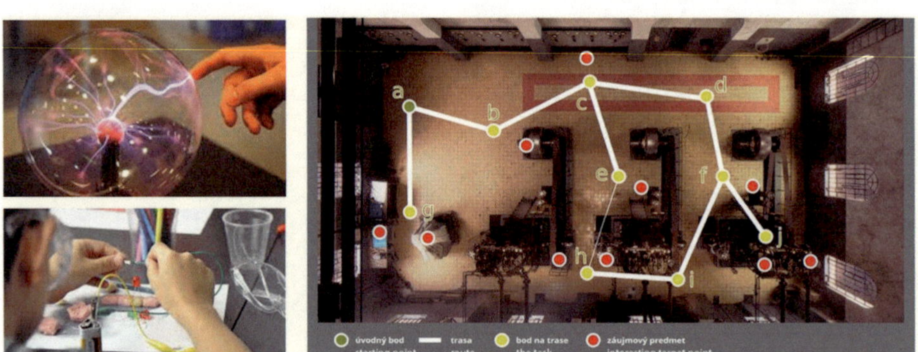

Figure 14.
Testing of classical manual education and haptic-virtual via VR (authors: FA STU, Živica, ZSE, a.s., iPARTNER s.r.o., CRATE, 2017).

- They are efficient, bringing clear and quick comprehending of the lost historical objects by different tools as 3D printing, VR, AR, or holographic models.
- Efficient non-formal haptic learning.

The installations using synchronized movement in VR and animated virtual elements induce immersive and subjective experiences. The presentation in old power plant Machinery Hall used not only audiovisual elements, but also it was supplemented by the real oil and diesel smell.

Supplementing elements from the real environment improve visitor's immersion in virtual space and his imagination. His potential to create long-term memories is also increased. In addition, the installation, which is presenting historical objects and spaces by similar methods, is also more attractive to younger, but also to older audience and is universally accessible by everyone.

As the installations using VR are attractive to wide public of every age and also to a professional public, the knowledge about historical and cultural values of the historical buildings and monuments is easier transferred and communicated. These immersive technologies proved to be efficient and appropriate tool for memorialize objects of the lost heritage and to reinterpret its importance for today and for the future (**Figure 15**).

By the mixed reality, the visitor is teleported into the virtual space with the ability of moving and viewing the space in a natural way. VR also allows people for disabilities to move through and to explore the space, without barriers, which would be not possible or too expensive otherwise.

In the case of old power plant in Piešťany, the HTC Vive showed to be less usable by people with motion disabilities than Oculus. The virtual space is usually perceived from the first-person view. This point of view could be also modified by using different perspectives (frog's or bird's perspective) and scales (the observer could be smaller in comparison to the model and vice versa).

The VR presentations offer the opportunities to experience the past, future, different fictions, or visions. Visual stimulation is supplemented with textures or materials from the real world as dust and smell, present in the old, preserved spaces and buildings, as the smell of the oil in the Machinery Hall.

Visits of the lost interior from 1906 of Machinery Hall in the old power plant in Pieštany are possible from anywhere, as VR with motion synchronization allows it. Synchronization of the real movement with the virtual one is convincing and validates the application of mixed reality as a tool for presentation of the lost industrial heritage in the contrast with its contemporary design, comparing these often very different states of the space [16].

Figure 15.
Picture of the virtual machinery hall with machine equipment—at the first stage of the power plant in 1906 and the haptic presentation in former machinery hall (3D model: O. Virág, M. Ganobjak, V. Hain; photo: V. Hain, 2017).

Figure 16.
Planned addition of haptic technologies—Microsoft HoloLens 2, interactive tablets for ARin the old power plant in Piestany (scheme: V. Hain, 2021).

7. Addition of new haptic technologies

Various studies researched the links between real and virtual by combination of various sensorial stimulations. To induce natural behavior in the spectator, the mixture of haptic and audio stimulations from real world and visual stimulations from virtual world was successfully used. The reason was that the spectator related the presented virtual space more easily to the real one.

The viewers perceived and comprehended the proportions and scale of the virtual elements more accurately as seeing them in the scaled physical 3D models or on the 2D displays as sketches or blueprints. In contrast, perception of scale was complicated, when visual stimulus from virtual environment was mixed together with visual and touch stimulus from the real environment on the scaled physical 3D model.

To achieve the more accurate perception of scale by the visitor in this combination of used stimulations from virtual and real environment, choosing a location in the scaled physical 3D model and then exploring it from that point of view in VR or by the camera would be more appropriate. This is implied in the described studies by the application of augmented haptic virtuality instead of using conventional augmented reality [17].

Therefore, the need arose to supplement the new available haptic technologies, which will be implemented in the premises of the Pieštany power plant in 2022 and subsequently their impact on users will be further tested (**Figure 16**).

The case study questions the relevance, meaningfulness, usability of VR, and its applications in entertainment. Some psychology researchers also indicate that improperly VR applications may lead to being isolated from the actual world that forced binocular imagery may cause brain disorder, and that its applications are not explored in the long-term view. In the described research, the VR is becoming a practical instrument for teaching wide public about lost historical objects. In comparison to various controversial applications of VR, this case study may be understood as appropriate and reasonable practical use of this technology [18].

8. Discussion

Using mixed reality (MR) as a tool for presentation and education of audience about industrial heritage is based on advanced technological skills in this area, but also to properly evaluate education level of the presentation. It requires to adapt the presentation to its targeted audience. The concept of using Haptic Technologies (HT) is not only the element of synergy, used in an organized complex design process, but in addition it is a crucial educational tool in MR.

Method that is trying to return the works to life can be called "virtual renewal." There are similar projects in the world often appear as "digital reconstruction" [19]. Virtual method recovery was in collaboration with students successfully tested even during the last pandemics semesters. For distance reasons, teaching students are able to study architectural works within reach your site and if the situation so allows, students can also verify the current state of the object in-situ via the "Urban Walk—Interactive planning method" [20] or the "Industrial Walk" in conjunction with the "Before after method" [21]. Only time will tell how successful it will be, but the growing development of haptic technologies is an important aspect for the future education.

9. Conclusion

With HT, the described case study has reimagined the industrial heritage history and brought something what was not possible to create physically to a present viewer. Learning about our lost historical objects is now easier and more accessible to wide public with this applied interactive technology. By focusing gaze on specific targets in scene, the interactive elements can be activated, and thus the user is informed and learnt by more natural way.

Visitor tracking is also a good educational element in understanding how people perceive the local industrial heritage sites, as much as they are interested in them and how to attract as many new participants as possible through HT.

For each experimental study of education about historical remains, the precise study of the subject is required. For that reason, the described case study will be used as a foundation for subsequent research of the HT applications in the education and preservation of industrial cultural heritage.

Acknowledgements

This project has been supported with public funds provided by the Slovak Arts Council FPU 16-362-03415 and project KEGA.

Author details

Vladimír Hain*, Roman Hajtmanek and Dušan Kočlík
Faculty of Architecture and Design, Slovak University of Technology in Bratislava, Slovakia

*Address all correspondence to: vladimir.hain@stuba.sk

References

[1] Bartolomei C, Ippolito A. Managing cultural heritage: Digital documentation and archiving. In: 2015 Digital Heritage. Granada, Spain: IEEE; 2015. pp. 259-266. DOI: 10.1109/DigitalHeritage.2015.7419506

[2] Hain V, Löffler R, Zajíček V. Interdisciplinary cooperation in the virtual presentation of industrial heritage development. Procedia Engineering. 2016;**161**:2030-2035. DOI: 10.1016/j.proeng. 2016.08.798. ISSN 1877-7058

[3] Ganobjak M. Virtual presentations as a tool to present values of extinct historical objects. In: Bardkontakt 2017. Proceedings 2017: Monuments and heritage sites in development programs of municipalities and regions. Bardejov: City Bardejov; 2017. pp. 181-188. ISBN 978-80-972776-7-3

[4] Sommerer C, Laurent M. Art as a living system: Interactive computer artworks. Leonardo. 1999;**32**(3):165-173. DOI: 10.1162/002409499553190

[5] Steed A. The virtuality continuum revisited. In: Grimshaw M, editor. The Oxford Handbook of Virtuality. New York: Oxford University Press; 2014. pp. 430-435. ISBN: 978-o-19-982616-2

[6] Keppl J, Špaček R. Latentné formotvorné činitele architektonického priestoru. In: Architektúra a urbanizmus. Bratislava: ÚSTARCH Slovak Academy of Sciences; 1986. pp. 237-252. 20 p. ISSN 0044-8680

[7] Hain V, Hajtmanek R. Mixed reality in the presentation of industrial heritage development. In: Sobota B, Cvetković D, editors. Mixed Reality and Three-Dimensional Computer Graphics. London, UK: IntechOpen; 2020. pp. 107-127. ISBN: 978-1-83962-622-7. DOI: 10.5772/intechopen.92645. Available from: https://www.intechopen.com/chapters/72350

[8] Turek I. Didaktika. IuraEdition. Bratislava: spol. s.r.o; 2008. 595 s. ISBN 978-80-8078-198-9

[9] Šimkovič V, Zajíček V, Hajtmanek R. User tracking in VR environment. In: Prokhorov S, editor. Proceedings - 2019 International Conference on Engineering Technologies and Computer Science. USA: IEEE; 2019. pp. 80-84. DOI: 10.1109/EnT.2019.00022. ISBN 978-1-7281-1915-1

[10] Basori AH, Daman D, Bade A, Sunar MS, Saari N. Proceedings of the 7th ACM SIGGRAPH International Conference on Virtual-Reality Continuum and Its Applications in Industry. New York, US: Association for Computing Machinery; 2008. pp. 1-2. 37 pages. DOI: 10.1145/1477862.1477910

[11] Robles-De-La-Torre G. Principles of haptic perception in virtual environments. In: Grunwald M, editor. Human Haptic Perception: Basics and Applications. Basel: Birkhäuser; 2008. DOI: 10.1007/978-3-7643-7612-3_30

[12] Hain V, Ganobjak M. Forgotten industrial heritage in virtual reality. Presence: Teleoperators and Virtual Environments. 2017;**26**(4):355-365. DOI: 10.1162/PRES_a_00309

[13] Vojteková E, Gregorová J, Polomová B, Sásiková K. Monument restoration—a controlled task does not limit creativity. World Transactions on Engineering and Technology Education. 2018;**16**(3):s.269-s.274. ISSN 1446-2257 (2018, 0.263 - SJR, Q2 - SJR Best Q). SCOPUS: 2-s2.0-85054991147

[14] Hain V, Hajtmanek R. Industrial heritage education and user tracking in virtual reality. In: Cvetković D, editor.

Virtual Reality. London: IntechOpen; 2019. pp. 45-65. DOI: 10.5772/intechopen.90679. ISBN 978-1-83880-861-7

[15] Sinčák P, Andrejková G. Neurónovési eteInžinierskyprístup (1.diel) [Internet]. 1996. Available from: http://www2.fiit.stuba.sk/~cernans/nn/nn_download/Sincak_Andrejkova_vol_1.pdf [Accessed: November 25, 2019]

[16] Hronský M, Kočlík D, Morávková K. Overlooked heritage: Interiors in Slovakia. ALFA. 2021;**2**:30-37. DOI: 10.2478/alfa-2021-0011. ISSN 2729-7640

[17] Salman A. Haptic Technology—Feedback, Devices, Working Principle, Applications. Available from: https://electricalfundablog.com/haptic-technology/ [Accessed: October 29, 2021]

[18] Guttentag DA. Virtual reality: Applications and implications for tourism. Tourism Management. 2010;**31**(5):637-651

[19] Spallone R. Reconstructive architectural and urban digital modelling. In: Mehdi K-P, editor. Advanced Methodologies and Technologies in Government and Society. Hershey, Pennsylvania, USA: IGI Global; 2019. p. 1. DOI: 10.4018/978-1-5225-7661-7.ch046

[20] Hanáček T, Fejo K, Hain V. Urban walk podmostie. In: Projekt. Roč. 60. Vol. 5. Bratislava: Spolok architektov Slovenska; 2018. pp. 10-15. ISSN 1335-2180

[21] Fejo K, Hanáček T. Elemental/E01: Základypremenymesta—scenárobytnejzóny. 1st ed. Bratislava, Slovensko: Spektrum STU; 2021. p. 56. 117 p. ISBN 978-80-227-5108-7